Allium

THE ORNAMENTAL ONIONS

Dilys Davies

*Published in association with
the Hardy Plant Society*

Timber Press · Portland, Oregon

First published 1992

© Dilys Davies 1992
Line illustrations by Graeme Robb

*Typeset by Rowland Phototypesetting Ltd,
Bury St Edmunds, Suffolk
and printed in Hong Kong*

*Published by
B.T. Batsford Ltd
4 Fitzhardinge Street
London W1H 0AH*

*A catalogue record for this book is available from the
British Library*

*First published in North America in 1992 by
Timber Press, Inc.
9999 S. W. Wilshire, Suite 124
Portland, Oregon 97225, USA*

Paperback edition printed 1993

ISBN 0 88192-241-2

B.T. Batsford Ltd
HARDY PLANT series

ORNAMENTAL GRASSES   *Roger Grounds*

MECONOPSIS   *James L.S. Cobb*

CAMPANULAS   *Peter Lewis and Margaret Lynch*

Forthcoming:

BORDER PINKS   *Richard Bird*

(Previous page:) *A. narcissiflorum* (Author)

The Hardy Plant Society

The Hardy Plant Society was formed to foster an interest in hardy
herbaceous plants. It gives its members information about the
wealth of both well-known and little-known hardy plants, how to
grow them well and where they may be obtained.

Members receive regular bulletins and newsletters and there are
regional and local groups, genus and special interest groups, a
seed exchange, plant sales and shows. The Society has a special
interest in plant conservation and has begun a programme of re-
introducing desirable plants from the past.

More information may be obtained from the General Secretary,
Hardy Plant Society, Bank Cottage, Great Comberton, Worcestershire WR10 3DP.

Contents

List of Colour Plates   4
List of Figures   5
Acknowledgements   6
Introduction   8
1 **Alliums in History**   12
  Ancient Western Civilisations   12
  Eastern Civilisations   15
  America   15
  Folklore and Medicine   16
2 **Edible Alliums**   20
  Recipes   26
3 **Botanical Classification**   29
4 **Geographical Grouping**   35
  Distribution   35
  Plant Introductions and Collectors   37
  Survey of Allium Bibliography   39
  Conservation   44
5 **Cultivation**   45
  Propagation   47
  Pests and Diseases   49
6 **Alliums in the Garden**   53
  Easy Alliums   53
  Alliums for the Rock Garden   55
  Alliums as Pot Plants   55
  Alliums in the Bulb Frame   56
  Alliums for the Border   57
  Alliums for Damp Sites   58
  Alliums to Avoid   59
  Colour Ranges   60
  Alliums as Cut Flowers   62
7 **A–Z of Selected Allium Species**   63
APPENDIX I   Section Definition According to *Flora Europaea* (1978)   148
APPENDIX II   Allium Collections   152
APPENDIX III Societies and Suppliers   153
APPENDIX IV   Hardiness Zone Maps   155
Select Bibliography   157
Glossary   158
General Index   163
Index of Species   000

Colour Plates

A. *narcissiflorum*	1	A. *flavum*	97
A. *paradoxum* var. *normale*	7	A. 'Globemaster'	99
A. *barszczewskii*	9	A. *goodingii*	100
A. *zebdanense*	11	A. *insubricum*	103
A. *moly*	13	A. *lineare*	105
A. *obliquum*	24	A. *moschatum*	108
A. *karataviense*	26	A. *neapolitanum*	111
A. *kharputense*	33	A. 'The Pearl'	112
A. *sphaerocephalon*	36	A. *olympicum*	115
A. *cyathophorum* var. *farreri*	38	A. *oreophilum*	116
A. *akaka*	40	A. *paniculatum*	118
A. *falcifolium*	46	A. *peninsulare*	120
Drumsticks in the border	50	A. *carinatum* ssp. *pulchellum*	
A. *oreophilum* var. *ostrowskianum*	54	'Album'	122
A. *schoenoprasum* 'Forescate'	54	A. *saxatile*	128
A. *pendulinum*	61	A. *scorodoprasum* ssp. *jajlae*	130
A. *acuminatum*	64	A. *scorzonerifolium*	130
A. *amabile*	67	A. *senescens*	132
A. *caeruleum*	75	A. *senescens* var. *glaucum*	134
A. *callimischon* ssp. *haemostictum*	76	A. *subhirsutum*	138
A. *christophii*	83	A. *unifolium*	143
A. *cyaneum*	89	A. *rubrovittatum*	151
A. *dichlamydeum*	92	A. *cyaneum*	152
A. *ericetorum*	94	A. *polyastrum*	162
A. *fistulosum*	95		

Figures

Bulb forms: bulb on rhizome; a single bulb 30
Cross sections of leaves 30
Cross sections of scapes 30
Buds with spathes 31
Perianth segments 31
Stamen formations 31
Forms of stigma 32
Structure of the ovary 32
Locules of ovaries 32
Various patterns of bulb coat 33
Variation in seed shape 33
Allium ampeloprasum var. *babingtonii* 68
Allium angulosum 24
Allium beesianum 71
Allium brevistylum 73
Allium callimischon ssp. *haemostictum* 56
Allium calocephalum 77
Allium campanulatum 77
Allium caput medusae 79
Allium cernuum 81
Allium cupanii 88
Allium cyathophorum var. *farreri* 90
Allium douglasii var. *douglasii* 92
Allium geyeri 16
Allium karataviense 104
Allium mirum 107
Allium moly 48
Allium oleraceum 114
Allium regelii 124
Allium robinsonii 125
Allium rupestre 127
Allium senescens subsp. *montanum* 133
Allium sikkimense 134
Allium subhirsutum 149
Allium textile 139
Allium thunbergii 140
Allium tricoccum 25
Allium triquetrum 27
Allium tuberosum 57
Allium unifolium 142
Allium ursinum 18
Allium validum 144
Allium victorialis 17
Allium vineale; A. carinatum; A. scorodoprasum 59

Acknowledgements

So many friends have been drawn into my involvement with alliums, but it was predominantly the influence of Marvin Black who, viewing with some amusement my initial collection, encouraged an increasing interest in those seedlings' fortunes. Not only did he pressurise me into plant photography, he made sure that enthusiasm burgeoned by introducing me to the superb country, magnificent flora and wide circle of friends in his beloved Pacific North West. Until then I had been most content with the modest fells of the English Lake District, despite their sheep-depleted pastures. Sadly he died in 1987, having increased public interest in plants in the wild and in gardens on both sides of the Atlantic.

American field trips invariably produced a few alliums when guided by such experts as Wayne Roderick, Panayoti Kelaidis and Coleman Leuthy, or organised by Dan and Evie Douglas. Alice Lauber and Sharon Collman have been indefatigable in directing me into the mountains, even when pouring rain has made plant hunting a mite uncomfortable.

Encouragement of Alliumophila in its many aspects has been directed by Dennis Thompson who found plant sites, obtained photocopies, arranged visits to Washington State, Oregon and Colorado, keyed out specimens or found coveted books in out-of-the-way bookshops.

Innumerable packs of seed or bulbs have arrived through the post. Ken and Gillian Beckett, Eleanor Fisher and Pat Davies have left few seed lists unscanned and have hunted out so many plants, references and photocopies for me. I have also a debt of thanks for all the help given by Brian Halliwell, John Fielding and Jerry Flintoff whose alliums are now on film.

Lawrie Springate has loaned material for further photography, given valued comments on the text, ransacked libraries and been always ready to discuss any aspect of *Allium* minutiae. Furthermore, his enthusiasm has been a great source of encouragement.

I am also grateful for plants, seeds, information and help from so many others in the quest for *Allium*. The list is incomplete, but must include: Colonel P.R. Adair, Alan Bloom, Felicity Baxter, Leonard Cama, Ray Cobb, Trevor Crosby, Peter Cunnington, David Frost, John Gregory, Dr Peter Hanelt, Richard Hann, David Haselgrove, Sheila Holborn, Rick Kyper, Roy Lancaster, Joanna Langhorne, Harry Lill, Mark McDonough, David Mowle, Godfrey O'Donnell, Bill Owen, M.R. Salmon, Roger Turner, John Watson, Rodger Whitlock, Rex Wilmshurst and Peter Yeo.

While many people acknowledge the help of their secretary, I have a great debt of gratitude to my partners Bob Button and Alan Wilkinson, who while they might not know an allium from an *Amaryllis* taught me how to use a word processor and willingly came to my rescue when inevitably I stalled the mechanism.

Graeme Robb has endured my bullying over the illustrations. Graham Rice has introduced me to publishing, while staff from Christopher Helm Ltd, A. & C. Black Ltd, and B.T. Batsford Ltd have all been involved with *Alliums*. I have been particularly grateful for the help of Robert Kirk.

The Royal Botanic Gardens, Kew, being involved in a taxonomic survey of Section *Allium*, have made it possible for Brian Mathew to proof read the species' entries. I am exceedingly grateful for his helpful comments and most appreciative of his fitting the reading into his very crowded calendar. *Allium* taxonomy being so confusing, the genus far from sparse in species, I have been fortunate in having such a knowledgeable guide.

Finally, while I am grateful for my family's long suffering over the last years' preoccupation, I must commend Tudor's patience. When he promised some years ago to endow me with his worldy goods, I guess he little thought this commitment would include sponsoring a work on onions.

A. paradoxum var. *normale* (Author)

Introduction

Dedicated *Allium* buffs can be counted on the hands, with a possible toe thrown in. This is a rather odd situation for, among the 800–1,000 possible species, there are many to captivate the botanically uninterested through the sheer floral pleasure they provide. Others pose problems of cultivation to challenge enthusiasts who revel in the difficulties of growing, and growing well, the most taxing of high alpines. Fears that *Allium* appearing on the show benches will waft a malodorous aroma towards the judges are, with few exceptions, unlikely.

The cerebral pleasure to be gained from a study of a complex genus appealed to Professor W.T. Stearn, the authority on *Allium*, who has confessed that he became interested in the genus early in his career, considering that it would provide sufficient content for a lifetime's study. One hopes his wealth of knowledge will be gathered together with all the resources that computers can now bring to scholarship.

As matters rest, the monograph of Regel compiled in 1875 remains the last comprehensive collection of *Allium* data. Long, tiring and expensive searching through any available flora is the only route to botanical information on species. This search is not eased if material is available only in Latin or Chinese. In Britain, unless one lives within easy reach of London or Edinburgh, research is doubly arduous, since provincial libraries are not all packed with specialist volumes. Photocopies help the search but one must know where to look and whom to ask for help in obtaining them.

By arrangement with the publisher, the Hardy Plant Society has embarked on a series of monographs to be written by members with a particular interest in a selected genus. These books are designed primarily with the practical gardener in mind rather than the student of taxonomy. Hopefully both groups may find a convenient collation of information in this small study.

In the case of *Allium* there are peculiar problems: firstly, the absence of collected information for the interested grower. *Flora USSR* published in 1935 describes 228 species, without counting synonyms. *Flora Europaea*, 1978, containing an Alliaceae section written by Professor Stearn lists 110, apart from synonyms. The later *Flora of Turkey* and *Flora Iranica* contain 141 and 139 species respectively, plus listings of ambiguous sightings. Further information is fragmented through numerous smaller local floras, many dating from the last century.

Secondly, to produce a mass of cultural detail for the gardener is hardly useful if he or she has little idea of the species being grown. Some of these difficulties were resolved by considering what sort of *Allium* literature I myself would find helpful both for identification and culture. To this end a list was compiled of plants available in commerce or from seed lists, whether commercial or from the plant societies.

In this context *The Seedlist Handbook*, originally compiled by the late Bernard Harkness, was a wonderful standby. It details seed appearing in the lists of the

A. barszczewskii (Author)

American Rock Garden Society, the Alpine Garden Society and the Scottish Rock Garden Club. In easily available format under each species, the common name, country of origin, height, colour and garden habitat are listed, together with primary bibliography. For many years this has been my garden bible: the second edition, now ragged and muddy, shares a corner of my greenhouse or garden seat. The third edition is still clean if well used and kept in some order for house use. A new edition has recently been published.

Other species included in my list were those I had grown, those I would hope to grow, those seen in the wild and those killed by lack of suitable cultivation. Some have been included as a warning to others not to grow them or, if grown, to recognise the potential hazard. Lastly, space has been allocated not necessarily on merit—undue proportion being given on occasion to some I just happen to like.

In *Allium* cultivation, as in other genera, a wide discrepancy is obvious between different gardens and gardeners. In 1971, we started cultivating 0.3 ha (¾ acre) of fellside in the English Lake District. The garden is almost vertical, which is just as well as the annual rainfall often attains a horrific 305 cm (120 ins). There are obviously wetter ecoclimates, but not many which gardeners try to cultivate. In the month of July 1988 alone, 66 cm (25 ins) of rain fell—hardly a warm dry summer. These conditions are not conducive to good growth of Dianthus, most silvers or many alliums. The soil is very shallow, short of humus, with a depressingly low pH. Ferns, liverworts and mosses do excessively well. So why grow alliums?

In the natural peat beds of the lower garden several of the Asiatic species grew quite happily. In a compulsive collector's need to find more, a great number of unsuitable species perished. Raised beds of stream grit in the open upper sites allowed wide experimentation. The absence of information and paucity of sources of supply

fired a project of assaying the hardiness of any species grown in the two areas available to me in Cumbria and Lancashire.

Seed came through the kindness of innumerable friends, from society lists, botanic gardens and from wild collections. As the pots of seedlings proliferated so the difficulties of identity increased. A nonbotanist has little idea where to rummage, but slowly the available literature was sifted through and a checklist of species and synonyms attempted.

Dabbling in *Allium* taxonomy would seem to pose problems even for the professional. There are so many species and the whole northern hemisphere to hunt through, it is not surprising that a recent monograph is not available. For an amateur gardener, the problems are even greater—both in taxonomy and in cultivation.

Well aware of the multitude of mistakes to be made in the gathering together of some of the *Allium* knowledge of others, I can only enter a plea of mitigation. If an outline of the vital statistics of some of the species, plus some knowledge of where to delve for further information, will be of use to others interested in the genus, then a modest compilation such as this may, I hope, have some value.

Quite deliberately, detailed botanical information, on stamen formation for example or chromosome counts, has been omitted as inappropriate in a book intended primarily for gardeners. Nor has differentiation into Sections been included, there being some lack of conformity between all authorities consulted. A surprising number of synonyms are listed because so many appear in seed lists or on *Allium* misnamed on sales tables.

Frequently my own experiences have been included even when it is quite obvious that cold wet gardens with excessive rainfall have little in common with gardening in warmer sites. I have risked my neck occasionally with some comments of my own, with which other growers may disagree.

After labouring on the species scheduled I found there were many more that I would still like to add. To the sceptics who may wonder whether 600 words might not be too many to expend on alliums, I find 60,000 leaves me with regrets that no more could be squeezed in.

Alliums are still a great fascination. An undeservedly neglected genus attracting a smallish circle of enthusiasts, plus the odd fanatic, has been a source of great enjoyment and most importantly the focus of friendships that I would otherwise have never known.

Hartsop, October 1989

A. zebdanense (Author)

1 *Alliums in History*

Almost as pervasive as the aroma of fried onions, reference to this adaptable genus can be recalled from the Ancient World. Across the whole northern hemisphere, probably since the first hot meal was prepared, alliums have provided the necessary flavour that turns plain fare into gourmet living.

The first of the onions are unlikely to have evolved before the time when early in Earth's history the land masses of the southern hemispheres broke away and migrated toward Antarctica. The known ability of bulbs to remain dormant under high temperatures might have saved plants in the hotter equatorial regions, allowing those borne into cooler latitudes to flourish in the Antipodes, though no endemic alliums are recorded there. So it would be reasonable to assume that early marsupials never had the pleasure of nibbling the pungent foliage of plants now so familiar to us.

From the beginning of time plants that have formed part of the food programme have evolved even under such unsophisticated culture with which we credit our distant ancestors. It may be possible that our attitudes are too patronising, that primitive man was perfectly capable of roguing poor strains and hybridising, if not with colchicine, with infinite patience in a world moving at a machineless slower tempo than today. With time to observe and cogitate, a crafty sense of expediency may have evolved a programme of plant breeding that would surprise us.

The origins of onion and garlic are lost in the ages, though the prototypes of onions and leeks appear to be found in Iran, Afghanistan and Central Asia, the hazy area that today is almost as unapproachable as in the days of Marco Polo and the spice routes. That they were highly prized as adding some of the spice and flavour to life is undeniable. Bread is bread but beer and onions add a dimension of flavour.

Ancient Western Civilisations

One of the earliest civilisations studied in the West was that of Egypt. Historian, amusing raconteur and recorder of myths, Herodotus travelled in Egypt some time around 450BC. His account of inscriptions, since destroyed, on the pyramid of Khufu (Cheops) built around 2500BC listed the amounts of radishes, onions and leeks provided for the labourers building the huge monument. Even allowing for exaggeration on the part of the translating pyramid guide, 1,600 talents of silver is an enormous sum for groceries.

Tomb inscriptions dating from this period of the Old Kingdom contain menus for the dead, with baskets of onions the sole vegetable. Lower down the social scale, the ploughman's lunch was even then bread, onions and cheese; or for high days and feasts, fish with a beaker of water.

Greek labourers fared slightly better, a skin of wine accompanied the bread and garlic. Horace, 65–8BC, no doubt disliking the aroma of the agricultural worker, deplored the addition of garlic, calling it more lethal than hemlock. In a letter to

Maecenas, he wrote: 'if it ever happened to you, my dear Maecenas, to taste such a flavouring, I pray the Gods that your Mistress will put her hand over your mouth and, refusing your kisses, will take refuge from your caresses under the bedclothes at the very end of the bed.' So much for Grecian garlic.

The Moly of Homer (*c.* 1000-800 BC, *Odyssey* Book X), hardly comes into this account of Alliaceae. Odysseus used Moly, a herb with milk-white flowers and black root to block the spells of Circe and thus avoid being changed into a swine. The plant in question might have been a Hellebore but hardly the yellow-petalled *A. moly* from the Iberian Peninsula. Tennyson followed this tale with The Lotus Eaters. If the accident-prone Greeks of the Odyssey were intent on delaying their eventual return home from the Trojan war then a bank of alliums would have been the very last thing on which to recline; tears to the eyes, yes, languishing amnesia, no.

Egypt was the granary of Ancient Rome. Juvenal the satirist, AD?60–?140 and Pliny the Younger, AD?62–?113, perhaps tongue in cheek, commented on the enthusiasm Egypt showed for garlic and onions, even to their elevation into gods, with oaths being taken on them.

Egyptian priests were forbidden to eat beans and other pulses, onions too but these were allowed to the gods. Paintings show their altars loaded with the bulbs or offered by priestly hands. The temples of Cybele, a Phrygian goddess whose worship spread throughout the Mediterranean,

A. moly (Author)

were closed against any with garlic-laden breath.

Manna cannot have been as appetising as we are lead to believe. An evocative passage from *Numbers* 11:5 (900–750BC), contains a complaint by the Children of Israel stranded in the Wilderness: 'We remember the fish we did eat in Egypt freely, the cucumbers and the melons and the leeks and the onions and the garlick. But now our soil is dried away, there is nothing at all, besides this manna, before our eyes.' To this day politicians still complain of the ingratitude of the liberated.

Both the Jews and the Egyptians were sophisticated people with a flexible written language and in the case of the Egyptians a talent for importing the best from distant lands. The Queen of Sheba led an exotic trade delegation from the land bordering the Arabian Sea. Quite probably alliums from the Indian subcontinent, Southern Persia or infiltrations from Central Asia, carried along the coast in ancestors of the Arabic dhow, formed part of her exotic baggage.

For centuries Rome was the centre of the civilised Western world. While Greece was the land of drama, philosophy and political hara-kiri, Rome was an empire of practicality. Hot sun, hard work and bread would be the peasants' lot, with wine and garlic for hard earned flavour. Doubtless the reeking crowds owed rather too much to garlic, today still an essential ingredient of Italian cooking.

ONION LANGUAGE

Even the names we use stem back to the Classical World. *Allium*, or *alium*, was the Latin name for Garlic. We find this root in Italian *aglio*, and in Spanish *ajo*. France uses *ail*. A very pungent garlic-saturated mayonnaise, *Aioli* (*Ali-oli* in Spanish), adds flavour to the dullest of salads. *Aillade* is the name given in the south of France to dishes flavoured with garlic sauces. By the time Linnaeus classified

plants in 1753, the term *allium* had appeared as a derivative from the vernacular.

While garlic had many cloves under its covering tunic, the onion was round, complete and architectural. To distinguish the two, the one bulb quite probably became *unio* and later *union*. Medieval spelling produces several forms and the French have adopted *oignon*.

In AD230, Apicius an epicurean writer appears to have considered onions rather less important than either leeks or garlic. The recipes he used utilised onions more as a seasoning than as an essential ingredient.

The Latin name for onion was *caepa* or *cepa* and from this we derive our species and section names. In modern Italian this appears as *cipollo* (note *Cipollino in agrodolee*, tiny onions in sweet-and-sour sauce), in Spanish as *cebolla*.

A further link with the past is the Greek *prason*, leek, from which we derive ampeloprasum, *ampelo* = vine, the allium that grows in vineyards. *Skorodon* was garlic and onion *krommuon*, names which now find use in the classification of alliums.

The Latin word for leek was *porrum*, which becomes in modern Italian *porro*, in French *poireau*, in Spanish *puerro*. Spain was part of the Roman empire until the invasion of the North African Moors introduced an alien culture. Indeed both Trajan and Hadrian, emperors of Rome, were of Spanish ancestry. The Spanish Onion was and is a major crop, so familiar that it became a feature in music hall songs.

The French leek is *poireau*. Breton onion sellers were a familiar sight before the Second World War along the south coast of England and Wales. Cultural exchange between these people has a venerable history even evolving language similarities in the old tongues of Breton, Cornish and Welsh.

Germanic people flowed through Europe, bringing the languages of Angles, Saxons and Jutes to the fringes of the erstwhile Roman Empire. From the Anglo-

Saxon root *leac*, all the following evolved: *leek*, English; *Lauch*, German; *luk*, Russian; *look*, Dutch. Garlic has derived from spear-herb: *gar* (spear), *leac* (plant or herb). *Leac-tun* was one of the Anglo-Saxon words for the kitchen garden. Not until the Norman Conquest did *onion* begin to replace *enneleac* in common usage.

Leeks thrive in the cold and damp of the British Isles. Introduced by the Romans, they persisted in the Celtic fringes of the Empire, as the invading Anglo-Saxon hordes swept into Britain forcing the natives into peripheral mountainous lands. Ironically the very name 'Welsh' derives from the word for foreigner—'welsche' in German. The Welsh onion derives from the same etymology.

The origins of a national interest in leeks by the Welsh seems untraceable. Perhaps the spear herb attained a mystical value in times when people equated appearance with potential attributes. Both leeks and daffodils are Welsh national emblems, appearing in the eighteenth-century writings of Iolo Morganwg. The association of the two widely differing plants may, however, stem from their sharing of a common name *cenhinen*, pl. *cenin*. The leek appears as the regimental badge of the Welsh Regiment of Foot Guards, formed in 1915 as part of the Brigade of Guards.

Several surnames have evolved from the genus. Onions, itself derived from Old French, oddly enough rarely means seller of onions. More frequently it appears as son of Ennion in the form O'Nions and pronounced oh-ny-ons. Garlic(k), Garlicke is Old English and, as one would expect, is the seller of garlic. Leek may be a seller of leeks but can also refer to the town of Leek (a stream or brook) in Staffordshire. The *Penguin Dictionary of Surnames* lists Ramsbotham, Ramsbottom, Ramsdale, Ramsden as place names stemming from Old English meaning Wild-garlic valley rather than rams' or ram's valley. In similar fashion Ramsey is the Wild-garlic island.

Eastern Civilisations

Few translations from the Chinese and Japanese are available so that we can compare their versions of the history of the Alliaceae. Many *Allium* species or their prototypes have, however, been cultivated century after century.

Garlic, *Suan*, was mentioned in the *Calendar of the Hsia*, a Chinese book dating two thousand years before the birth of Christ.

From the Far East, *A. tuberosum* Chinese Chives, *kin ts'ai* has come to us as a food crop that still retains identity with wild-collected species.

America

The North American Indian certainly made alliums part of his food pattern but as a chance finding rather than a cultivated crop. Most of the wild species seem to have been articles of diet. *A. geyeri*, found in the Koosskooskia River area of Idaho, was known as *Omoir* by the Nez Perce Indians, while *AA. reticulatum, canadense* and *cernuum* are all listed as dietary items.

A degree of botanical skill was required by the gatherers of allium bulbs. In many parts of the country, certainly in eastern Washington State, *Zigadenus venenosus*, the Death Camas, grew in the same localities and looked not dissimilar. As its common name implies, a mistaken bulb could prove fatal. If today we smile, not too many modern gardeners would walk into the woods and with assurance pick any of the native fungi for dinner.

The *Journals of Lewis and Clark* have plenty to say of the wild onions' use as commercial assets. One entry details their value when over-indulgence in the gassy bulbs of edible Camas (*Camassia leichtlinii* and *C. quamash*) brought on attacks of colic. The two explorers used them also in poultices, for sick Indians as well as for their own men.

Allium geyeri

By 1775 Romans was describing the cultivation of both garlic and leeks by the Choctaw Indians (*Natural History of Florida*, 1: 84, 115).

The Okanagan-Colville Indians of Washington and British Columbia harvested *A. cernuum* between April and June, shortly before the plants flowered. The onions were rolled on mats, removing soil and the outer skins, while the leaves were plaited. In some areas they were eaten raw, in others steamed in pits overnight. They could be eaten there and then or dried and strung for winter stocks, later being soaked in water overnight to rehydrate them. Other tribes made pressed cakes of the cooked bulbs before drying them.

A. douglasii and probably *A. geyeri* were also dried and cooked in a pit but mixed with lichens or Black Camas for flavouring.

Nitinaht Indians on Vancouver Island seemed not to have eaten the local *A. cernuum* until Captain James Cook sailed into Nootka Sound in 1778, finding the 'wild garlick' in plenty. Once they realised the sailors were willing to buy the bulbs, harvesting was conducted briskly.

Local species with small bulbs hardly would have seemed worth cultivating. Following the introduction of onions, leeks and garlic from Europe the native peoples, like Candide, went forth to cultivate their gardens. Recognition of the American alliums as decorative species has waited in the main until the present century. Even today the many native species are largely unappreciated.

Folklore and Medicine

A separate volume could be written on the place that alliums have occupied in folklore and herbal practice, both ancient and modern.

Leaving the Ancient World and passing into Mediaeval times is mainly a matter of written records. When the barbarian hordes overran the tidy Roman world, scholarship was kept alive in the Celtic fringes of north-west Europe, in isolated monasteries and in the oral tradition of country people.

Witchcraft became a potent and feared force during the Middle Ages in Europe. With 'wise women' acting as healers, spells and herbs became inextricably mixed. Modern medicine with all the resources of laboratory mystique still recognises the placebo effect, a potent force even today in our ostensibly non-credulous society.

The value of garlic as a counter-attack in the fight against vampirism is curiously difficult to establish. *A. victorialis* was considered potent against an indeterminate

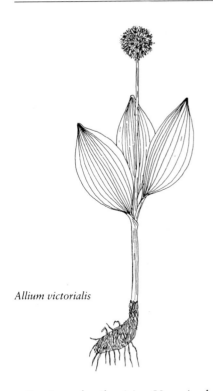

Allium victorialis

neuralgia and certain earaches. Certainly it gained a reputation during the First World War as an antiseptic. As for plague, the Four Thieves' Vinegar, redolent of garlic, was believed to give immunity to corpse robbers in their chosen profession.

Yet folklore often contains more than a grain of truth. The *Encyclopedia* carries a list of Gilbertian patter, antibiotic, antirheumatic, antimalarial, antihelminthic, hypotensive, hypoglycaemic, rubefacient, intestinal disinfectant, carminative, balsamic and efficient remover of corns.

Recent medical research, in New York and India, has been conducted with scientific methodology. The American studies have found significant beneficial effects from allium medication in cardiovascular disease. However, toxic effects from overdosage should be considered before anyone decides to live on a primarily onion/garlic diet. These studies comment that fresh garlic is most efficacious, while proprietary or dried products are not as potent. Eating parsley is recomended as a deodorant.

The Indian studies suggest that the smellier ingredients of garlic—the sulphur-containing fractions are the most potent—may be helpful in reducing fats in the blood (perhaps frying the garlic would defeat the purpose?). Both these studies advocate further research.

More mundane usage has garlic chopped with lard or vaseline then rubbed on the chest for whooping cough or bronchitis; against leprosy—and acne; mixed with honey and eaten for a rheumatism remedy; or, incredibly, marinated in red-wine vinegar then left to mature in sunshine for use as a toilet water!

Galen, practising in Rome in the first century, described garlic as the countryman's best antidote and is believed to have prescribed it frequently. As for veterinary practice, a good dose was advocated to purge a dog of worms.

By Elizabethan times, the term garlic-eater denoted low social status. Later be-

collection of evil spirits. Vampire legends were current in ancient times but modern mythology was apparently founded on the history of Vlad the Impaler who terrorised fifteenth-century Hungary. Gaining impetus with the publication of *Count Dracula* by Bram Stoker in 1897, the market was flooded in this century with a spate of horror films. All we learn from this fantasy is assurance that garlic flowers attached to doors and windows repel vampires.

Such a ubiquitous and potently smelly plant was bound to attract a powerful charisma. The *Macdonald Encyclopedia of Medicinal Plants* offers an incredible list of uses, ranging from insect repellent to glue. We are also told that chickens will lay more eggs when garlic is a food additive, and it has been credited with delaying the rotting of stored fruit and deterring Weevils from invading grain stores.

Medical uses from the same source include alleviation of toothache, trigeminal coming an expression of xenophobia, garlic eating has only returned to popularity in Britain in recent years. Few Europeans today would consider a steak even adequately cooked without the essential savour of garlic.

Chives have been recommended as a wash for mildew on cucumbers and gooseberries, and for planting near apple trees to prevent scab. On the farm they can be used in mash for rearing young turkeys and chickens. Freeze drying is the best method of preservation for kitchen use.

Ramsons, *A. ursinum*, can be used medicinally in place of garlic. Surprisingly few gardeners, overrun by the plants, consider eating them as a means of eradication, but the bulbs are edible and the young leaves can be used in salad. Slimming diets have included juice of Wood Garlic.

'Onions can make even heirs and widows weep,' wrote Benjamin Franklin. Peeling onions under a running tap, heating small onions before peeling, holding a crust of bread or a toothpick in the mouth or using patent choppers have all been advocated to check tears. Wearing glasses and a bath cap also help, at least the cook's hair won't smell later. Oddly enough, the tears that follow onions are not inclined to block the nose or cause reddening of the eyes in the manner of emotional tears. Why not?

Deodorising of hands has been attempted by rubbing with chopped parsley, lemon, celery leaves or salt before the final wash with soap and water. Breath sweeteners include milk and black coffee. Likewise chewing parsley, sugar, coffee beans or a clove are said to be effective. Commercial deodorant tablets are also a good safeguard. Strangely, a chopped onion on a plate takes away the awful smell of modern gloss paints.

Medicinally, everything that garlic can do, so can onion. Once more the *Macdonald Encyclopedia* piques one's curiosity. Onions can repel moths and mosquitoes, prevent rust, impart a gleam to glass or copper, remove freckles, promote hair growth, improve hearing and allay hysterical attacks of several types, induced quite probably by all this strange activity – truly a wonderful plant, almost too good to be eaten and adding only about 125 calories to the day's rations.

According to French folklore, onions may be recommended to weather forecasters, a race of people somewhat prone to error.

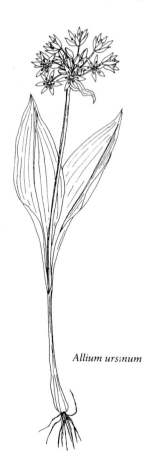

Allium ursinum

Si l'oignon n'a qu'une pelure,
L'hiver passera sans froidure.

When the onions grow thin skins,
Winter cold won't freeze your shins.
                    (free English translation)

Good Friday is a day to avoid planting onions, for according to a Brazilian legend they will not grow. Around the world are numerous American Onion Creeks. There is the Onion River in Vermont, while Burgundy has the river L'Oignon. Onion domes proliferate on Central European church domes and Asian mosques, with the Cathedral of the Assumption in the Kremlin, Moscow supporting some of the most justly famous. While the device is common in architecture, onion representation is rare on pottery.

In my home town of Preston, Lancashire, onion skins have been used for dyeing. The town crest is the Paschal Lamb and Cross of St John the Baptist, associated with the early days of St Wilfred, AD634–709/710. According to tradition every Easter Monday, Pace Eggs (from Latin—*pascha*—Passover, Easter) are rolled down the banks of the park overlooking the River Ribble. These are hard-boiled hens' eggs, decorated according to the resources and skill of the rollers. After wrapping the eggs in onions skins tied with cotton, boiling produces an attractive brown satiny pattern on the shells: one of the age-old methods of colouring used before the invention of felt tip pens. Rolling boiled eggs down a slope is not a very exciting pastime in the age of television. A few more years will probably see the last of the Pace eggs.

With justice perhaps, R.L. Stevenson called the onion 'The rose of roots'.

Smell and taste in alliums arise from the presence of diallyl disulphide ($C_6H_{10}S_2$) or diallyl trisulphide ($C_6H_{10}S_3$). Research into the chemical and physiological properties of these compounds is being pursued in several academic centres. Coronary heart disease, high blood pressure, raised fats in the blood and factors that influence clotting are all subjects of investigation.

Toxicity is also under review. The effect on the eyesight of long-term ingestion of *A. scorodoprasum*, the Rocambole or Sand Leek, is mentioned by Li Shih-Chen in *Chinese Medical Herbs*. Ordinarily, a very high intake of allium is required before ill effects occur.

Excretion is partly via the skin, accounting for the miasma that may envelop the garlic eater as well as for the unmistakable halitosis. This odour appears more pronounced after eating fried alliums than when the plants are cooked in water, despite the incredible smell that boiling garlic creates in a kitchen. Surprisingly few foods or drugs appear to be dispersed via perspiration. Should the value of the genus in skin conditions be shown, this trait could be a therapeutic blessing.

2 Edible Alliums

There is little need to stress the importance of the genus among world crops. Over the ages, selection of the more valuable plant material for kitchen use has resulted in some confusion in provenance. Not only the geographical origin but the original species is often impossible to assess accurately. Furthermore, many modern books on vegetable gardening confuse still further and conflicting accounts of the nature of Scallions, Tree Onions, Egyptian Onions or Potato Onions abound.

Being neither a competent vegetable grower nor a willing cook my knowledge of the commercial alliums is limited. This short compilation will, I hope, contain the more feasible attributions.

SHALLOT, ESCHALOT

'Eat the white shallots sent from Megara.' Ovid, *Ars Amatoria*

The mild flavour of the Shallot has made them favourites for pickling, for flavouring stews and soups or for eating raw in salads or sprinkled over meats. In AD230, Apicius published several recipes for their use.

Martial, AD40–104, the Roman poet and author of the *Epigrams*, had some interesting comments to make on the properties of the species.

Alexander the Great was said to have found the Shallot in Phoenicia and introduced it into Greece.

The *Askalonium krommoon* was described by Theophrastus, fourth century BC, and the *Cepa ascolonia* by Pliny the Elder, author of the 37 volume *Historia Naturalis*, who claimed that it originated in Ascalon in Syria. Many of the early botanists repeat this statement, but Alphonse de Candolle (A.DC) writing in 1885 was inclined to think the Shallot a form of *Allium cepa*. Michaud in his *History of the Crusades*, 1853, claimed its introduction by returning warriors.

Shallots were grown in the gardens of Charlemagne along with leeks, chives and onions. About 75 herbs were listed in the *Capitulare de Villis Imperialis*, AD812. Lusitanus, in 1554, provides evidence of widespread cultivation in Europe but Gerard, in 1597, does not mention them. Shallots for American gardens were listed by McMahon by 1806.

The original *A. ascalonicum* Linnaeus was based on *A. hierochuntinum*, a Palestinian species described by Boissier. Shallots are now considered a variant of *A. cepa*.

CULTIVATION Bulbs are planted at the end of February with the tips just protruding above the ground, then lifted and dried at the end of July. The smaller bulbs are saved as stock for the following year.

ONION
A. cepa Linnaeus

From the earliest recorded times, onions have been important. They have been cultivated for so long that their origins are obscure. *A. oschanini* B. Fedtschenko (*A. cepa* var. *sylvestre* Regel) has been nominated as an ancestor. An Eastern origin is certain, but whether from Iran or India or those lands sweeping eastward towards China is impossible to say; certainly it is no longer found growing wild.

Alexander the Great is credited with their introduction into Greece, having found them in Egypt. They became a regular part of his armies' rations, having been thought excellent for exciting martial ardour.

Following very ancient references from Egypt, onions were mentioned by Hippocrates in 430BC, while Theophrastus listed at least four local variants in 322BC. Dioscorides, in AD60 limited his descriptions to yellow or white, round or long, while in AD42 Columella wrote of the Mariscam called by countrymen *unionem*. Pliny the Elder, AD79, commended the round onion, commenting on the stronger flavour of the red varieties.

References to these ubiquitous vegetables appear thick and fast until, by the seventeenth-century, varieties are classified by shape, size, colour and country of origin. An account at this time mentions that the Roman colonies during the time of Agrippa grew eight pound monsters of a Russian variety, worthy forerunners of the 'footballs' to be seen in flower shows today.

From the East, legend recalls how Satan left the garden of Eden after the fall of man. From the spot where he planted his right foot, onions sprang up, from his left footprint came garlic.

Alphonse de Candolle (A.DC) writing in 1855 of America refers to *xonocatl*, which Humboldt had said was used by Amerindians, and that Cortés had seen garlic, leeks and onions on the army's march to Tenochtitlan. He was doubtful though of their identity with European vegetables known by these names. Other writers claim Peruvian onions and garlic had arrived there from Europe, originating among herbs sown by Columbus in 1494 on Isabela Island.

Onions were cultivated in New England early in the seventeenth century. In 1779 General Sullivan destroyed Indian crops including onions near Geneva, New York State. A gardening publication of 1863 listed 14 varieties.

Around the same date at least 60 varieties of all shapes, sizes and colours were grown in France.

CULTIVATION Onions require rich, well-drained soil and a sunny site to flourish. Seed may be sown in the greenhouse in January, then planted out in April. Planting sets is common practice with the crops being harvested in late August. Bending over the tops in the middle of the month hastens ripening. Store in a frost-free place.

EGYPTIAN TREE ONION
A. Cepa 'Viviparum'

Dalechamp recorded in 1587 his great surprise at seeing small bulbs growing in an onion head in place of seed.

A degree of confusion surrounds the identity of the plant, commonly found in kitchen gardens, with largish bulbils formed in the flower head, flowers usually being absent. The plants are almost evergreen, the top-heavy apices bending over towards the soil surface and allowing the bulbils to root and continue growing while still attached to the parent plant. The bulbils are excellent for pickling and for use in stews, having thin skins, though the inner scales are stained purple at the bases.

CULTIVATION Perfectly frost hardy, the bulbils may be harvested throughout the winter and are most conveniently grown in single rows; a thicker planting becomes untidily congested through the plants' self-rooting habit.

A. canadense L. has much the same habit, being designated the Tree- or Bulb-bearing onion (Egyptian Onion), by J.C. Loudon in *Horticulture*, 661 in 1860.

The *RHS Dictionary* offers several suggestions for Tree or Egyptian Onion: *A.*

cepa var. *bulbiferum*, with bulbils among the flowers, *A. c.* var. *proliferum*, few-flowered with large leaf-bearing bulbs in the inflorescence, and finally *A. c.* var. *aggregatum*, which bears bulbs at the apex of the scape.

Contrariwise *Hortus III* lists the Aggregatum group, (which includes *A. cepa* vars. *aggregatum*, *solaninum*, *multiplicans*) as Potato Onion, Multiplier Onion and Ever-ready Onion. The Proliferum group includes Tree Onion, Top Onion and Catawissa Onion. All thoroughly confusing.

LEEK
A. porrum   Linnaeus

Another vegetable that has been cultivated from the earliest times and not naturally found in the wild. *A. ampeloprasum* was probably the prototype of the leek.

The Romans split *porrum* into two forms: *capitatum*, the leek proper, and *sectile*, the form we call chives. Pliny the Elder wrote that the best leeks came from Egypt or Aricia in Italy.

Rumours circulated that Nero regularly ate them, hoping to improve his singing voice, thereby attracting the nickname Porrophagus, the leek eater.

Medieval Europe was well acquainted with the vegetable but, while their introduction to England is given as 1562, this is almost certainly incorrect. Monastic cultivation would have been current much earlier.

Traditionally the leek has been a Welsh emblem. Captain Fluellen forced Pistol to eat not only his words but plant, skin and all, in scene 1, Act V of *Henry V*. Indeed, Shakespeare based this whole confrontation on the national importance of an otherwise insignificant vegetable. It seems impossible to find an earlier reference. A vague association with St David, a sixth-century victory over the Saxons when leeks were worn to distinguish them from the enemy, or Welsh participation in either Crecy or Poitiers have all been suggested.

Welsh soldiers fought under the Black Prince and Edward the Third at Crecy in 1346 and were almost certainly present at Poitiers in 1356; but the chroniclers make no mention of the leek.

Unlike onions, leeks are mentioned by Gerard. In 1596 he published a catalogue of plants and the following year *The Herball*, a rewrite of *Stirpium Historiae Pemptades Sex* of 1583 by Rembert Dodoens.

In the New World, Cortés noted leeks growing and by 1775 the Choctaw Indians were cultivating them.

At the present day, leek growing is taken very seriously in north-east England; competitive, giant vegetables are raised to astonish the general public.

CULTIVATION Seed is sown early March–mid April and transplanted when 15 cm (6 ins) high. Deep dibber holes are made in the bed and the transplants watered in. Soil can be banked against the leek sides as plants grow, the crop being harvested as required. The perfectly hardy plants are ideal winter vegetables, even in cold areas.

GARLIC
A. sativum   Linnaeus

Originating probably in Central Asia, garlic has been known from earliest times, spreading through the Far East, Europe and North Africa. The precursor is thought to be *A. longicuspis* Regel. Believed to have been introduced into China almost from the dawn of history, references are reported from 2000BC. More certain mention occurs in later Chinese literature from the fifteenth century.

Dioscorides used the name *skorodon hemeron* and the *allium* of Pliny was thought to be garlic. The Turkish word is *sarmisat*, *sarmisat disi* being the cloves.

Garlic was available in England before 1548 and earliest references in America mention that Cortés ate it in Mexico, probably imported by the Spaniards.

CULTIVATION In England garlic is not often seen in flower, needing a mild climate for perennial culture. Commercially sets are offered, rarely seed, being planted February to April and harvested August.

CHIVES
A. schoenoprasum Linnaeus

Chives are found throughout the north temperate zone, even in Alaska and Siberia. Quite possibly the food plant with the most pleasing flower, there are very good forms to grace the kitchen garden.

Gerard described them thus: 'a pleasant Sawce and good Pot-Herb', Charlemagne held the same opinion for chives were listed in the 75 or so herbs to be grown in his garden.

Moore in *Baileya* comments that chives turn up in many guises—a long list follows: AA. *ampeloprasum, angustoprasum, darwasicum, frigidum, ledebourianum, libani, oleraceum, scaposum, tibeticum, wallichianum, winklerianum* and two unlikely cultivars A. 'Siskiyou Mt' and A. 'Ruby Gem'. Caveat emptor.

CULTIVATION Culture is remarkably easy, division of the slender bulbs being all that is required.

Other Alliums Used as Food Sources

Most alliums are considered to be edible though palatability obviously varies. Doubtless any plant not obviously poisonous has formed, however temporarily, part of the food supply, with wild colonies supplementing main crop alliums. Professor Stearn relates how he sampled all available *Allium* species during the Second World War on behalf of a Ministry of Food disturbed by the absence of imported garlic. Most species qualify for occasional use with the following finding a corner in the literature.

A. acuminatum
Makes palatable grazing for sheep and cattle but does not grow thickly enough to be economically important. *A. brevistylum* is also relished by the same animals, at least when young and succulent.

A. akaka
Used as a vegetable in Iran, either on its own or with rice in a pilau.

A. ampeloprasum
Eaten raw in certain parts of southern Europe. Variously known as Great-headed Garlic, Wild Leek or Levant Garlic.

A. angulosum
Grown in Siberia as Mouse Garlic or *mischeitschesnok*, usually salted for winter use.

A. canadense
A source of food for the American Indian and early settlers, has been superseded as a crop.

A. cernuum Wild Onion
Found growing from western New York to Wisconsin and southwards. In 1674 with *A. canadense* formed almost the main source of food for the explorer Marquette and his group as they travelled from Green Bay to the present site of Chicago, then a porterage site. The city was first settled in 1830, the name being Indian, translated by some as powerful, by others as skunk or wild onion. In 1878 the prairies around Chicago were described as an unbroken carpet of delicate pink onions. The bees' honey from this harvest was said to lose

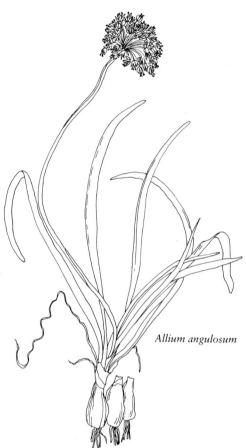

Allium angulosum

A. oleraceum Field Garlic

Records from Sweden include the flavouring of soups and stews with the young leaves, or fried with other vegetables.

A. reticulatum

The roots formed part of the diet of North American Indians. This species is now a synonym of *A. textile*.

A. roseum Rosy Flowered Garlic

Has edible roots; eaten in Mediterranean countries.

A. rotundum (A. scorodoprasum subsp. rotundum)

From Europe and Asia Minor. The Greeks living in the Crimea were reported to have used the leaves as food during the last century.

A. obliquum in Preston, Lancashire (Author)

any allium flavour by the ventilation produced in the hives as the bees fanned their wings. Allium honey was considered quite the equal to that gleaned from clover.

American Indians also made *A. cernuum* part of their diet.

A. chinense

Has been, indeed still is, an important vegetable crop in China and Japan; likewise *A. tuberosum*, Chinese Chives and *A. fistulosum*, Japanese Bunching Onion or Welsh Onion.

A. obliquum

Found in Siberia, was used as a substitute for garlic and cultivated on the lands flanking the River Tobol.

A. rubellum

Found in Europe, Siberia and Central Asia. The leaves are dried and used as seasoning by the hill people of India, the bulbs also being eaten.

A. senescens

Widespread through Europe, Siberia and the Far East, it has been eaten as a vegetable in Japan.

A. sphaerocephalon Round-headed Garlic

From early times this was eaten by people living in the region of Lake Baikal, Siberia.

A. stellatum

Was part of the food supply of dwellers in North America, both the indigenous population and the early settlers.

A. tricoccum Ramps

Like its European counterpart, Ramps grow in eastern North America in woods and on hillsides, being traditionally gathered for seasonal feasts.

A. ursinum Bear's Garlic, Ramsons, Buckrams, Gypsy Onion, Hog's Garlic.

The many folk names bear witness to its ubiquity throughout Europe and northern Asia. In 1597 Gerard reported that the leaves were eaten in Holland. The bulbs were boiled and also used as salad in England. As a pot-herb the leaves are rather pungent. In Russia, through to Kamchatka, the bulbs were gathered as winter food. Despite the many woods that have large colonies of the plants, few people today gather either leaves or bulbs.

A. vineale

A pernicious weed throughout Europe and naturalised in America, its former occasional use is reflected in the folk names of Crow Garlic, Field Garlic and Stag's Garlic.

Allium tricoccum

According to the US Department of Agriculture, cattle and sheep relish most onions but horses are not so enthusiastic. Unfortunately too much allium flavours cows' milk, so grazing must be selective. Elk in Yellowstone Park graze on the tops, particularly in spring, while bears dig up the bulbs.

Alliums for the Pot

Not many plant books carry a chapter of recipes but then not many genera are both decorative and edible. Out of the whole genus only plants of the Melanocrommyn section are said to be unpleasant to eat. This group includes *A. karataviense*.

The spirit of scientific enquiry has so far

A. karataviense (Author)

not included a research project that involves the author in eating as well as trying to grow every allium available.

'let him eat of onions and he will be doughty enough.' Martial

Apicius, a Roman writer, lived in the time of Augustus and Tiberius. This vegetable dish comes from his cookery book *De Re Culinaria* printed for the first time in Milan in 1498. Contemporary with Seneca, Pliny the Elder, Juvenal and Martial, and a great gourmet and glutton, Seneca wrote of him that when he slid into debt and down to his last ten million sesterces (= ¾ ton of gold), he committed suicide rather than face a life of penury unable to support his greed for food.

Liquamen or garum, a common flavouring, the Roman equivalent of bottled sauce, was made from 2 tablespoons of red wine, 3 oz salt, 3 anchovies and a teaspoon of dried origan (marjoram). Having boiled the ingredients together for 10 minutes the strained liquid was stored for use. Factories producing liquamen sprang up in Pompeii, Leptis Magna in North Africa and Antibes.

Plain salt could be used as a substitute.

'Wel loved he garleek, oynons and eek leekes. And for to drinken strong wyn, reed as blood.'

So runs The Summoner's Tale, from the Prologue to *The Canterbury Tales*. Chaucer, a poet and public servant, had some uncommonly aristocratic relatives. His wife was the sister of John of Gaunt's mistress Katherine Swynford. Katherine eventually married her Duke of Lancaster after the death of his second wife. Chaucer himself held an honoured place in the service not only of Edward III and the later usurping Henry IV but was particularly favoured by Richard II, a cultured monarch more interested in poetry, fashion, food and music than survival politics.

From the master cooks of this ill-fated king comes a manuscript entitled *Forme of Cury* (medieval for cookery), compiled around 1390.

CYMA ET PORRI
(Cabbage and Leeks)

Cook 1 lb cabbage and ½ lb leeks in boiling salt water for 10 minutes, then drain. Place the cabbage on an oven dish and pile the leeks over the top. The following mixture is poured over the vegetables before baking for 20 minutes.

　　1 tablespoon each of oil and wine
　　2 teaspoon each of ground cumin and
　　　caraway seeds
　　¼ teaspoon ground coriander and
　　　pepper
　　1 teaspoon of liquamen

CABOCHES IN POTAGE

Take caboches and quarter ham and seeth hem in gode [good] broth, with oynonns y mynced and the whyte of lekes y slyt and corve smale, and do [add] thereto safronn and salt and force it with powder douce [allspice].

'Onions . . . the juice anointed upon a pied or bald head in the Sun, bringeth the haire againe very speedily'. Gerard

The recipe following might do just that.

AIOLI FROM TOULON

For five people take 8 cloves of garlic pounded into a fine paste with salt. Gradually add the yolks of 2 eggs, then slowly drip in 3 oz of olive oil and the juice of half a lemon, as if making mayonnaise. The resultant sauce should be very thick and is delicious with jacket potatoes, almost any vegetable, snails or fish.

Balzac enjoyed onions, even for breakfast, considering them excellent for the mind, 'rendering it subtle and putting to flight base notions and prejudices'.

Maybe not quite so early in the day! However, chopped chives go very well stirred into scrambled eggs. Or try the eggs on garlic toast.

'Ay, leeks is good.' Scene 1, Act V, *Henry V*

Uncommonly tasty with a flavour in a class far outranking its title is

LEEK AND POTATO PEASANT SOUP

Fry 2 chopped leeks in 2 tablespoons of butter in a covered pan for 15 minutes. Add 2 potatoes cut in chunks, 2 pints of water, salt and pepper and simmer for a further 30 minutes. Puree the vegetables, check seasoning and serve.

'Onions make thin . . . stamped with Salt, Rue and Honey and so applied, they are good against the biting of a mad dog'. Gerard

Cornish tin miners survived hard lives of poverty and near starvation in the late nineteenth Century. Thousands emigrated to California, Australia, South Africa and Peru. Even in recent times the Folk Lore Society of Colorado held a Cornish dinner on the Eve of St Piran, the patron Saint of tinners. From these impoverished days comes

KIDDLEY BROTH OR KETTLE BROTH

A handful of *A. triquetrum* leaves, chopped squares of bread, dripping or bacon rinds (possibly butter on special days), pepper, salt and a few marigold flower heads (*Calendula*) were steeped in boiling water.

This was poor food for working men, though slightly better fare was served on holidays.

LIKKY OR LEEK PIE

Simmer 12 chopped leeks for just 5 minutes in 5 oz milk, salt and pepper, then remove with a slotted spoon and layer with

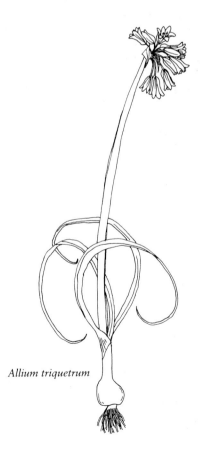

Allium triquetrum

8 oz sliced, unsmoked bacon in a pie dish. Pour the milk once more over the mix and cover the dish with an 8 oz topping of shortcrust pastry before baking for ½ hour at 350°F, 180°C, gas Mark 4.

Having taken the dish from the oven, ease off the crust and remove some of the liquid. Beat 2 egg yolks with 5 oz cream, fold in 2 stiffly beaten egg whites, then pile over the contents of the pie dish. Re-cover with the crust and return to the oven for a further 10 minutes.

Finally, a personal favourite:

SWEET AND SOUR BAKED ONIONS

Simmer 6 medium-sized peeled onions for 10 minutes, drain and save the onion water. Place the onions in a casserole, season with salt and pepper and sprinkle with 2 oz soft brown sugar. Dissolve a stock cube in the onion water and make up to 10 oz with water, pour over the onions.

Cover and bake at 350°F, 180°C, gas Mark 4, for 25–30 minutes until tender. Drain once more and, after blending 1 tablespoon of cornflour and 1 tablespoon of wine vinegar, stir in the onion liquid. Bring to the boil, cook 2–3 minutes, season and pour over the onions.

And for dessert? A recent festival in Gilroy, California marketed 'Garlic Topped Ice Cream'. A philosophical ten-year-old found this awful but not as bad as he had expected!

To drink with all these dishes? Only a real enthusiast would wish to serve

VODKA WITH GARLIC

Peel 3 garlic cloves, cut them in halves and soak in a bottle of vodka for 36–48 hours before serving chilled.

'The juice of Alliums who fondly sips,
To kiss the fair must close his lips.'

Anon

3 Botanical Classification

However much the practical gardener would like to enjoy plants without bothering with botanical detail, the moment arrives when he or she needs to know the correct name of the occupant of rock garden, pot or border. Pink Daisy, Blue Campanula won't do for very long; albeit with reluctance, horticultural Graeco-Latin must be used.

The *Allium* grower has as many or more problems than most. Five hundred, 600, 700—the list still grows, with synonyms as thick as bulbils all around. Not even the family name has been definitive until quite recently. A vast genus lost in a vaster family, the Liliaceae.

Why ubiquitous and most familiar vegetables should have a botanical background so confused is not a little intriguing. Plants that appear on the dinner table are not so very mysterious. But then what market trader would place his leeks or onions in a crate with lilies, strange bedfellows indeed.

Because the family of Liliaceae embraces plants with six stamens and a superior ovary—the base of the remaining flower parts are placed below the ovary, style and stigma—*Allium* found their place here.

In 1935, writing in his *Families of Flowering Plants*, Hutchinson argued that because the flowers were carried in a bracted umbel at the apex of a leafless stem, *Allium* should be placed in the *Amaryllis* family, with *Narcissus* and *Galanthus*. The inferior ovary of the Amaryllidaceae was considered to be less important than other factors.

Chemotaxonomic studies however indicated that while *Allium* shared some characteristic chemical features of Liliaceae, they lacked the alkaloids demonstrated in all the genera of Amaryllidaceae.

A study of Rust fungi on *Allium*, conducted by Savile in 1962, suggested that *Allium* were related to but more primitive than *Scilla* and related genera, and therefore should form a separate family of Alliaceae. This new family could be considered related to the immediate ancestors of Amaryllidaceae and Liliaceae.

Both chromosome and embryological studies a little earlier had not supported the transfer of *Allium* to Amaryllidaceae, so there the matter rested until the publication of *The Families of the Monocotyledons* by Dahlgren, Clifford and Yeo in 1985. A division of Liliaceae into 25 smaller families based on this work was adopted by the Royal Botanic Gardens, Kew. Consequently *Allium* are now found with 31 other genera including *Agapanthus*, *Brodiaea* and *Triteleia* in Alliaceae. *Nectaroscordum* and *Nothoscordum* are also included in the new family.

So what is an *Allium*?

Descriptions can be found in floras, heading the sections that contain the entries relating to *Allium*. The language is precise, botanical and couched in technical terms.

Included in the Glossary are many terms which may not appear in this account, but do recur in the various floras. Over the years compiling my own glossary has made botanical understanding far easier, for even some apparently simple term, like widespread, may have a qualified meaning. While this swells the list to daunting proportions, bringing together the language of alliums may smooth the path of under-standing.

According to *Flora Europaea*, an *Allium* may be described as 'a perennial bulb, distinguished usually by a characteristic smell of garlic or onion. The bulbs may be single or grow in clusters, while the leaves are threadlike or ovate, with a basal sheath'. Some species have slender bulbs, little more than the enlarged bases of the leaves, attached to a distinct rhizome which is a swollen rootstock. A remnant of rhizome may be found included in most species if only as a piece of the bulb base.

Allium leaves can be completely basal, short, as long or longer than the flowering stem, erect or lying limply on the ground. The basal sheath may lie below the surface or present fan-like at ground level.

In other species the basal sheath may clothe the lower portion of the stem, giving the impression of tiered leaves.

Great variation occurs in leaf structure: solid, hollow or the two surfaces may be concave-convex. Distinct ribbing of the central vein of the lower surface may form a keel, in others the upper surface may be channelled.

Cross sections of scapes

Thin, threadlike leaves, again some solid, others hollow, typify several species. Long hollow blowpipe-like leaves occur in several Onion relatives. Comparable to some Tulip species there are alliums with extra broad, glaucous, wavy-edged, even colour-tinted, leaves. Most of these are species from the Middle East; some are very decorative plants for rock garden or alpine house.

Many alliums have leaves that have withered by flowering time, rather spoiling their garden efficacy. In others the leaf tips have browned or yellowed by maturity. In contrast there are many that are almost evergreen, retaining fresh foliage throughout the season; these are mainly plants from areas of high rainfall and cooler summers. Their leaves may be util-ised as salad in their country of origin. Tiny teeth occur along some leaf margins.

Like the leaves, stems may be solid or hollow, tall or short, with below-ground portions.

Botanically the stems are scapes, having no leaves produced along their length. In those sheathed for half their stem, the leaves are produced basally with the sheath continuing upward. Some species with hol-low stems show variations in the diameter along the stalk length. A few species have scapes that are strongly ribbed, some are flattened into distinct faces, others slightly winged. Solid stems may be smooth, glistening, rough or glaucous. A few spe-cies have stem, leaves, flower stalks and buds, or any combination, markedly glaucous or bearing a delicate bloom.

The buds are initially enclosed in a spathe which may split into one, two or more valves, to reveal the flowers in a terminal umbel. The valves may be markedly uneven, strongly veined and dif-ference can occur in the persistence with

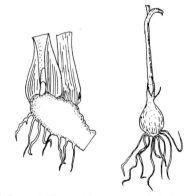

Bulb forms: bulbs on rhizome; a single bulb

Cross sections of leaves

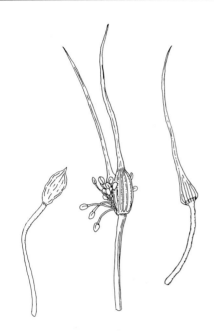

Examples of buds with spathes

which the valves remain below the flower head as the season continues.

Umbels are floral umbrellas, the flower stalks or pedicels being the spokes. While these arise from a central point, the stalks may vary enormously in length between species and also in an individual umbel. This characteristic is constant for any individual species. The pedicels may also be green, white, pink, lilac, purple, buff and any shade in between, sharing or contrasting with the perianth colour. Small bracteoles are often found at the pedicel base.

In dicotyledons the coloured petals are usually easy to distinguish, being very often the characteristic that gives the flower its appeal as a garden plant. Some dicots, to confuse, put all their vivid colour into bracts or sepals, e.g. *Euphorbia. Hacquetia* and *Cornus.*

Monocotyledons may have coloured flower parts that are not classified as petals, in the case of *Allium* these are known as tepals. At first sight this looks uncommonly like a misprint or an anagram. The flower head or inflorescence

may be referred to as an umbel, while the small individual flowers are sometimes called florets in the species descriptions in Chapter 7. Strictly this term is botanically incorrect, florets being the flowers of composites and grasses, but has been used to try to make the species' descriptions less obscure. A more correct term for the tepals would be perianth segments.

Segment shape, size, colour, texture and angle all contribute to identity. In some species the tepals rapidly wither, in others remaining stiff longer after seed has formed. Some excellent alliums can be grown for dried flower arrangements.

Flora Europaea defines the six perianth segments as being persistent, free or slightly joined by their bases and having one to three veins.

The structure of the six *Allium* stamens is very important in distinguishing species. While some may be joined at their bases forming a ring, others are free. Yet again there are stamens that are adherent to the tepal behind them. A further distinguishing feature may be extraordinary teeth

Perianth segments

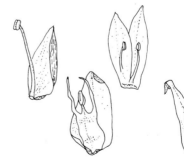

Stamen formations

growing alongside the filaments and some-times dwarfing them. The details may re-quire a hand lens or low power microscope to differentiate. In some authorities, for example *Flora Iranica*, there are clear drawings of the structure.

Again colour may be variable, or the filament length. In some species the stamens are included within the flower, that is they are as long as or even shorter than the tepals. In others the stamens are exserted, their length being greater than the tepals. The anthers, which face inward, may be differently coloured from the per-ianth or indeed the filaments which they cap. Occasionally pollen may also be a different shade.

Most gardeners will be content to ignore the fine details of stamen structure. In the description of species, only a simplified indication of stamen specification has been included. Those who require greater dif-

Locules of ovaries

ferentiation should find the bibliography will list an adequate reference.

Stigma, style and ovary form a carpel, the female seed-bearing unit of a flower. With the exception of some species bearing sterile flowers, each flower in a typical umbel carries a single carpel. The stigma at the top of the style may be simple, cap-headed or 3-lobed. In shade it may contrast or share the flower colour.

The ovary often shares the flower tend-ency to divide into three parts, separating or merely appearing lobed; very prominent in *A. kharputense* and *A. nigrum*. Each ovary has three locules which contain the ovules, the potential seeds before fertilisa-tion. Usually each locule carries two ovules, occasionally one, rarely three or more. Small crests and protuberances can be found on some ovaries. Nectaries are yet another distinguishing feature, lying in hollows between the locules either at the base or in varyingly shaped pits that can extend up the ovary wall.

Another important distinguishing fea-ture in *Allium* taxonomy is the structure of the bulb coat. Some have glossy, shiny, dry skins like the domestic onion. On market stalls there are red, white, green, buff, purple and muddy varieties. While for the cook the skin colour can be merely decor-ative the unfortunate botanist needs to pack a magnifying glass, for without one identifying *Allium* may be impossible. A network of fine patterns on the tunic may be the vital key. Diamonds, stripes, squares in the hotchpotch of fibres around the mature bulbs may be the only clue to iden-tity.

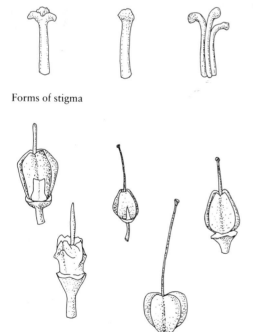

Forms of stigma

Structure of the ovary: various formations

A. kharputense (Author)

Various patterns of bulb coat

Variation in seed shape

Many of the North American alliums rely on the patterns of bulb coat formation for exact identification; further, most produce small bulblets.

Bulbs are a plant's way of circumventing the scorching droughts of summer. The winter rain or snow melt allows early growth of leaves and flowers. Fertilisation achieved, the plant sets seed; then, duty done and immortality hopefully attained, the parent retires underground. Above, a scorching sun may parch the land, the dead husked remains scatter and below ground the bulb stores all resources for the following season. In general terms, the larger the bulb, the more reasonable to suppose that the species originated in an inhospitable desert climate. A big store cupboard will more profitably see the season through.

In contrast, alliums from damper climates where the seasons do not differ so radically between rainfall and desperate drought, have become plants with smaller, slender bulbs, often clustered on a short rhizome or underground root extension. Without searching scrutiny some bulbs from areas of high rainfall appear to be merely fibrous roots. These plants tend to flourish happily in damp gardens far from their native habitat. In these species identification rarely devolves on minutiae of bulb tunics.

In the past little information has been available from the literature on the structure of seeds. Some have wrinkled skins, others smooth and shiny. Seeds may be spherical, wedge-shaped, almost flattened, black and varying in size. There may be netted patterns on the surface, small rough prominences or tiny warts.

Chromosome counts are being evaluated with increasing frequency but this is of academic interest to anyone outside an institution with laboratory facilities. Inter-relationships of theoretical importance can be determined by such research, meanwhile everyday assessment of good garden potential lies in simpler tests. Does it grow and is it worth garden space?

Keys can be produced using the difference in species for the identification of plants. Unfortunately, in a genus the size of *Allium*, with such a wide geographic spread, keys are useful mainly when the country of origin of a specimen is known. Plants grown from a seed list may have originated in Timbuctu but with a label that specifies Nevada. In such circumstances, inspired guess work and midnight oil will be necessary. Careful examination of all the plant parts, and a working knowledge of *Allium* variations should make possible the placing of the unknown into a defined Section.

Unfortunately, not all authorities agree on the criteria for a Section, nor are Section names uniform. There are also differences in the descriptions of plants common to separate geographical flora.

A simplified scheme of the Section division from *Flora Europaea* (see Appendix 1) demonstrates the parameters by which differentiation is attempted. Because of the disparity between Section allocation however, no attempt has been made in the species' descriptions in Chapter 7 to assign plants to a such an arrangement.

4 *Geographical Grouping*

Estimates of the number of *Allium* species around the world have ranged between 500 and 700. Geographically the genus is found throughout the northern hemisphere. With merely a handful of exceptions, species are absent from the southern hemisphere, unless artificially introduced.

Bulbs being a plant's method of withstanding drought, it is not surprising that the areas of the world where water is limited, or absent for at least one season, are areas of high *Allium* concentration.

During a discussion of the RHS Lily Group in 1967, some comparative figures were quoted.

For the whole of the USSR, Vvedensky (1963) estimated there were 233 species, with the Soviet Republic that borders Afghanistan in the north-east showing 15 endemics out of 80 species. Afghanistan itself boasts around 50, including 15 endemics. Other Central Asian areas included West Pakistan with 35, and 18 from the central and eastern Himalaya.

Iran has about 15 endemics among 60-odd species, Iraq 32 with nine endemics. Estimates for Turkey included 20 endemics among a further 60 species.

Figures for Syria and Lebanon, 40, were among those estimated by Naomi Feinbrun in 1948, with 25 in Israel, seven being endemic.

The Index to the *Flora Reipublicae Popularis Sinicae* (1980) carries 99 entries. Despite the stream of plants from Japan, Ohwi's *Flora Japan* printed in 1965 merely lists eleven *Allium*.

From North America an enumeration published in 1940 in *Herbertia* by C.V. Morton listed 97.

Europe may contain around 120 species.

While the tides of war lap across the continents and countries become no-go areas, estimates become even more difficult to confirm despite modern chromosome count techniques. The propensity of botanists to lump or split might also alter an *Allium* count by tens or hundreds in a decade. Fortunately *Allium* are not inclined to interbreed with much enthusiasm, for there are quite sufficient species in the half of the world in which they grow to cause sufficient problems of identity.

Distribution

Comparatively few *Allium* are native to the British Isles. Hardly surprisingly in a country with a high rainfall and low sunshine level, few bulbs have to take shelter from a blistering summer drought.

The *Concise British Flora* of W. Keble Martin, 1965 lists *AA. ampeloprasum, babingtonii, carinatum, oleraceum, paradoxum, roseum, schoenoprasum, scorodoprasum, sibiricum, sphaerocephalon, triquetrum, ursinum* and *vineale*. Many of these may well be introductions, in the wake of the Roman occupation in the 1st to 4th centuries as well as later more peaceful invasions via the cooking pot.

Around the Mediterranean littoral, *alliums* revel in hot, dry, gravelly regions. Ample springtime moisture followed by arid summer baking is relished by most of the genus. Cool, wet winters may allow growth to continue with bulb dormancy mainly confined to summer. On higher

A. sphaerocephalon (Brian Mathew)

ground, snow cover in winter provides cold weather protection for the bulbs. The warmer, wetter winters of north-west Europe are less suitable. The broad-leaved alliums, *AA. ursinum, triquetrum* and *victorialis*, are found in damper situations than the majority of their co-genors, in upland meadows, damp woods or streamsides.

Alliums from the Mediterranean zone have found homes in coastal areas of North Africa; there are, however, very few in the southern hemisphere. Don describes *A. rubicundum* and *A. synnotii* (?syn. *A. ampeloprasum*) from the Cape of Good Hope, South Africa; *A. gageanum* is another species from the same area (*Cape Flora Catalogue*, 1984).

Greece and the Greek Islands have numerous endemic species, for example *A. olympicum* and *A. callimischon*, small bulbs found in rocky, mostly free-draining outcrops. Moving eastwards, Turkey and countries still politically open to plant ex-

ploration continue to surprise with a stream of new species. The *Flora Turkey*, and *Flora Iranica*, tantalise with the description of plants still not easily acquired. 'Drumstick' alliums shoot upwards as the freezing gritty environment melts in a new year, very probably to end in nomadic pots and pans. Neat, decorative alliums for non-culinary, alpine house pots, such as *A. akaka* with encircling grey-green leaves much admired by Western gardeners, may find their way into Persian markets to garnish sheep's eyes.

Central Asia most probably is the ancient home of garlic, that currently sophisticated herb of today's Western culinary hedonist; but so many *Allium* listed in the *Flora of the USSR* are continually under review.

A. caesium and *A. caeruleum* have evolved a life style in salt marshes but the majority of *alliums* are found in areas where summers are hot and dry. Rain or snow in winter is followed by spring-flowering before the baking heat of summer has to be endured deep below the earth's crust.

Among the many alliums from Central Asia may be listed some of the tall drumsticks—*AA. giganteum, elatum, stipitatum, aflatunense* and *rosenbachianum*. Forms of *A. oreophilum* grow well on the rock garden, while dry borders sport *A. christophii* and *A. karataviense*.

Towards Nepal, Tibet and China an interesting trend appears; the pink, lavender and white flowers, ubiquitous in the desert-type alliums, begin to show more change in colour. Most of the blue alliums valuable in the Western rock garden appear. The wetter environment, particularly at higher altitudes, gives rise to fibrous roots and slender bulbs, with grass-like foliage still viable at flowering time. True blues and bluish-purple flowers are found flowering in late summer and autumn. *AA. beesianum, cyaneum* and *sikkimense* bring the loveliest of sky reflections to our gardens, to complement the shades of heaven from gentian or meconopsis.

Political exclusion has not dampened the Victorian and Edwardian fascination with China. Plant collections across the continent attracted some of the most illustrious, obsessive and inspired travellers. Wilson, Fortune, Farrer, Forrest and Purdom laid a foundation of plant hunting that is unlikely to be slaked for decades as slowly the trickle of Western visitors returns. *AA. forrestii, beesianum, polyastrum, cyathophorum* var. *farreri* are some introductions from this era.

Dry winters and wet summers mainly characterise the Himalayas and China, with alteration mediated by altitude. The climate of Japan tends to wetness year-long, with high summer rainfall. Alliums in these conditions once more produce small bulbs with rather fibrous roots, many flourishing in the climate of Great Britain. *A. virgunculae* and *A. thunbergii* are two alliums also well suited to alpine house culture.

North America has a rich heritage of alliums. The north-east has colder winters and hotter summers than European lands of similar latitudes. *AA. tricoccum, cernuum* and *canadense* range along the eastern prairies. From the Mid-West grasslands come species that include *A. stellatum*.

The majority of American alliums grow along the Pacific Coast lands and in the Rockies. California's climate, Mediterranean in nature, has produced a wealth of bulbs. Highly decorative are *AA. unifolium, campanulatum* and *dichlamydeum*. The Pacific North-West, coastal Oregon, Washington and western Canada has Western Europe climatic style. The great chains of mountains form the vertebrae of the continent, nurturing many alliums. From the northern reaches come *AA. crenulatum, brevistylum, geyeri, cernuum* and *acuminatum*; the list of species follows the ranges running down to Colorado, New Mexico and Texas, joining the alliums of the plains and drier grasslands.

Even within these sketchy outlines of *Allium* distribution there are great variations in climate and geological patterns. An excellent description of bulb distribution geography is to be found in *Growing Bulbs*, Martyn Rix (1983).

Plant Introductions and Collectors

'The vast family of the Garlics, ranging all across the northern and southern regions of the temperate zone, is disqualified in the garden on account of its prevalent and odious stink, which, combined with the predominance in the family of inconspicuous weeds, sheds a disability on even the beautiful species that here and there occur.'

Reginald Farrer,
The English Rock Garden

Such comments are naturally hurtful to any right minded *Allium* grower. One should remember, however, that Farrer like many newspaper reporters made statements for effect with some obscuring of the truth. While these enliven his prose, there have been rather too many rock gardeners who have taken every word attributed to the great man as directions straight from Heaven.

Following this broadside his three pages of *Allium* species are almost all favourable; censorious in the cases of *AA. ursinum, libani, victorialis* and *aschersonii*. Despite this prejudice, Farrer and Purdom introduced *A. cyathophorum* var. *farreri* and *A. purdomii*, the latter later identified as *A. cyaneum*.

Throughout the Far East many of the early explorers and plant collectors were missionaries. Later expeditions in the nineteenth century were financed by seed merchants and nurserymen from Great Britain.

George Forrest has been mentioned in

A. *cyathophorum* var. *farreri* in Hartsop, Cumbria
(Author)

connection with AA. *polyastrum, bee-sianum* and *forrestii* from Western China. A. *amabile* was another of his finds.

Van Tubergen, the Dutch firm, sent collectors into Turkey and Central Asia, introducing A. *christophii* into general cultivation circa 1903.

Albert, son of Eduard von Regel, and the Fedtschenko family—Aleksei, Olga and their son Boris—all collected in Central Asia in the region of Turkestan. A. *cyaneum* from Gansu (Kansu) is associated with the Russian botanist Przewalski. Maximowicz worked in Siberia. All four have *Allium* named after them.

In 1884 Surgeon-Major Aitchison, working on a border commission, as a sideline surveyed the flora of Afghanistan, and is commemorated in A. *aitchisonii*.

A. *mairei* celebrates its discovery in 1860 by Edouard Maire.

Collectors in America include David Douglas (1799–1834), a Scot after whom is named A. *douglasii*, and Thomas Nuttall (1786–1859), A. *nuttallii*. Carl Purdy of California sold native bulbs, mainly wild collected. The firm continued to operate from 1893 until the 1950s, being recalled by A. *fimbriatum* var. *purdyi*.

After the end of the First World War, interest in bulb growing increased but with the emphasis shifting from the professional collector to the amateur explorer. The activities of the Royal Horticultural Society (RHS) and the Alpine Garden Society in England added impetus to enthusiasm. A. *akaka* was sent back from Iran by E.K. Balls, an English emigré to California.

Following the Second World War, the botanist Peter Davis collected widely in the Middle East, editing the *Flora Turkey*. Oleg Polunin, an English schoolmaster, collector and writer travelled in the Himalayas and the Balkans. Patrick Synge, bulb grower and writer associated with the RHS, sought plants in Turkey and Iran. His friends and fellow seekers were Rear Admiral and Mrs Paul Furse; between them they explored as far as Afghanistan. Their wanderings were catalogued in the *Journals of the RHS*, accompanied often by the beautiful watercolours executed by Admiral Furse.

Flora Iranica contains many new *Allium* species described by the Norwegian Professor Per Wendelbo, who also botanised throughout Afghanistan.

Present day collectors appear frequently in lists of *Allium* species labelled by the initials of the expedition members. This list may be helpful in deciphering society seed lists, though it is not comprehensive.

JCA	J.C. Archibald	FB	F. Baxter
CDB	C.D. Brickell	PC	P. Cunnington
PF	P. Furse	BH	B. Halliwell
RL	R. Lancaster	EMR	E.M. Rix
WR	W. Roderick	JMW	J.M. Watson
EKB	E.K. Balls	KAB	K.A. Beckett
PC	P. Christian	PHD	P.H. Davis
CGW	C. Grey-Wilson	SVH	S.V. Horton
BM	B. Mathew	MACP	J. MacPhail
PMS	P.M. Synge		

Some collectors will be listed with their surname initials only in expedition codings.

Survey of Allium Bibliography

Alliums of Europe and Asia Minor

If the history of *Allium* covers almost the whole recorded history of Mankind, then a review of the bibliography traces the origins of scientific classification of plants.

Leaving the medieval world of spells, simples, herbal medicine and cookery, some of the earliest illustrations of *Allium* species appeared in the works of Brunfels, 1489–1534, Fuchs 1501–1566 and Mattioli 1500–1577. Their plants were Leek, Garlic, Onion, Ramsons and the weedy Crow Garlic.

In England *Allium* featured among others in the works of John Gerard, whose *Herball* was published in 1597 and John Parkinson, *Paradisi in Sole Paradisus Terrestris*, 1629.

The publication of *Rariorum Plantarum Historia* in 1601 opens the age of modern plant scholarship. Charles de L'Ecluse 1526–1609, was born in Arras, travelled and worked all over Europe before settling in Leyden, where he was later given a Professorship of the University. According to the convention of the age when all academic work was written in Latin, he is better known as Carolus Clusius. An accomplished botanist with an analytical eye, the woodcuts illustrating *Allium* in his celebrated work are clearly recognisable today, encumbered as they are by pre-Linnaean terminology. One species illustrated under the title of *Moly minus* was later cited by Retzius who renamed it *A. clusianum* in commemoration.

The genus *Allium* as an entity seems to have been initiated by the Swiss Albrecht von Haller, 1708–1777, the first botanist to collect all the extant alliaceous plants into one grouping. Previous writers having scattered onion, garlic and leeks through separate genera, Haller published in 1745 the monograph, *De Allii Genere naturali*

Libellus. Haller was primarily an anatomist, also pursuing his interests in poetry and botany with equal distinction.

The year 1753 was a turning point in the study of botany. The Swede, Carl von Linne, 1707–1778, revolutionised the nomenclature and classification of plants and animals. The diffuse, archaic, polysynthetic descriptions were transformed into the binomial system of today. Linne's epic work *Species Plantarum* was published in 1753 and included descriptions of 31 *Allium*. For the range of his knowledge of plants it is only necessary to skim through the list of species in a later chapter to see the name Linnaeus on almost every page. In any flora of the northern hemisphere, by general usage the initial L. standing alone is sufficient to denote the work of this giant of biological history. Yet even the innovative Linnaeus recognised and absorbed some of the earlier work of Clusius and Haller into his *Allium* methodology.

A Monograph of the Genus Allium was published in 1827. This was the work of George Don the younger, 1798–1856, and had originally been read in Edinburgh to the Wernerian Natural History Society the previous year. It is also available in Vol. 6: 1–102 (1832) in the Memoirs of the Society. He described 139 species, mainly in Latin, and some of his divisions of the genus into Sections are still accepted today.

Eduard von Regel, 1815–1892, working in Leningrad, published *Alliorum adhuc cognitorum Monographia* in 1875. Alexei and Olga Fedtschenko, having made plant collections during their explorations of Turkestan, had gathered species not previously described in the literature. Eventually Regel, having worked on the collections, produced a monograph of the whole genus, which also included several related genera. The species' descriptions and the key are in Latin, the notes and introduction in German. This and a later supplement *Allii Species Asiae Centralis* 1887 were published in *Acta Horti Petropolitani* (St Petersburg).

The Swiss botanist Edmond Boissier, 1810–1885, published an account in Latin of *Allium* species found in the region running from the eastern Mediterranean to the Indian border, in his *Flora Orientalis* (published in two parts in 1882 and 1884), Vol. 5.

Flora USSR under the editorship of Komarov appeared in 1935, the *Allium* entries in Vol. 4 being contributed by A. Vvedensky. Two hundred and twenty-five species of *Allium* from the European regions of Russia through Central Asia are catalogued, many having been originally described by Vvedensky himself. An English translation from the Russian was prepared by H.K. Airy Shaw, appearing in *Herbertia*, Vol. 11 in 1944, with a number of line drawings.

A useful work, despite inevitable name changes, containing the description of approximately 115 species and listing innumerable synonyms, was published by C.H. Grey in his three volumes of *Hardy Bulbs* (1937–8). Grey adds some practical details of cultivation and his personal evaluation of garden worthiness. The general style of writing and the format makes this an accessible work for those who find the compact information of a flora somewhat indigestible.

Flora Europaea, ed. T.G. Tutin (1978), contains the account of *Allium* contributed by Professor W.T. Stearn. In Vol. 5, pp. 49–70, 110 species are described, further synonyms being listed in the text. This work is essential reading for the study of European *Allium*.

Flora Iranica, ed. K.H. Rechinger 76: 1–100 (1971), contains descriptions of 139 *Allium* by the late Professor Per Wendelbo, many species having been first described by himself. The contribution on Alliaceae is published separately and contains both photographs and line drawings. The descriptive text is in Latin with notes in English.

Flora Turkey, ed. P.H. Davis (1965), with the Alliaceae contributed by F.

A. akaka growing wild in east Turkey (Brian Mathew)

Kollmann, describes 141 species in English with an appendix containing doubtfully recorded species. There are numerous maps illustrating geographical distribution, plus diagrams of stamen/tepal formation.

The European Garden Flora, ed. S.M. Walters, Vol. 1, 1986, contains entries for 96 *Allium* written by W.T. Stearn assisted by E. Campbell. The title of course relates to plants found in gardens in Europe without restriction to the country of origin, for the Alliaceae section also describes Asian and American species.

In Christopher Grey-Wilson & Brian Mathew's *Bulbs. The Bulbous Plants of Europe and Their Allies*, 1981, the chapter on *Allium* describes in 20 pages a multitude of species grouped according to their relationships, and illustrated with line drawings of flower shapes. There are also coloured drawings by Marjorie Blamey depicting 28 species.

Early volumes containing illustrated plates of *Allium* include *Flora Graeca*, J. Sibthorp and J.E. Smith. Sibthorp, 1785–1796, a professor of botany in Oxford, travelled extensively in the Aegean and eastern Mediterranean. The *Allium* plates were painted by Franz Bauer, 1758–1840, from live specimens as they were collected.

Pierre Joseph Redouté, known mainly by his Rose studies, contributed 35 plates of *Allium* to *Les Liliacées* (1802–1816), the texts being written severally by Augustin P. de Candolle (DC), de la Roche and Delile. The *Allium* illustrated were those currently available in France. *A. carolinianum* was mistakenly named at this time. Seen growing in a garden in Paris and believed to have originated in America, it was ascribed by De Candolle to the wrong continent.

H.G. Ludwig Reichenbach published *Icones Florae Germaniae et Helveticae* in 1848. The coloured plates illustrate very many of the European *Allium*.

Curtis's *Botanical Magazine*, published between 1787 and 1948, included approximately 60 hand-coloured plates of *Allium*, allowing for the inclusion of synonyms. The *Kew Magazine* took its place in 1948, substituting lithographs and increased text; only a few illustrations are listed in the Index to date.

The Bulb Book of Martyn Rix and Roger Phillips contains over 800 photographs of hardy bulbs, including about 45 *Allium* portraits. This is a book essential to almost any gardener, not just an *Allium* collector. One hopes that a companion volume covering bulbs from the Far East, and another featuring American species, will also be published.

There are of course many general works that feature a limited number of *Allium*. Those listed below offer a choice for the general reader. The specialist articles are to be recommended to the more committed *Allium* grower.

A FURTHER SELECTION

* = includes illustrations of *Allium*

Anderson, E.B. (1959) *Dwarf Bulbs for the Rock Garden*
——(1959) *Rock Gardens*, Penguin Handbook
——(1964) *Hardy Bulbs*, Vol. 1, Penguin Handbook
——(1966) *'Alliums for the Garden'*, *RHS Lily Year Book*, 29
Blanchard, K.S. (1970) *'The Ornamental Onions'*, *Journal RHS*, XCV
Chittenden, F.J. (ed.) (1956) *RHS Dictionary of Gardening*; and *Supplement* (1969)
Feinbrun, N. (1943) *'Allium Sectio Porrum of Palestine and the Neighbouring Countries'*, *Palestine J. of Botany*, Vol. III, no. 1
Fitter & Fitter (1974) *Wild Flowers of Britain and Northern Europe*
Hay, R. & Synge, P.M. (1969) *Illustrated Dictionary of Garden Plants*
Grey-Wilson, C. (1979) *The Alpine Flowers of Britain & Europe*
Janka, V. De (1886) *Key to the Alliums of Europe*, Budapest; trans. W.T. Stearn, *Herbertia*, 11: 1944
Kollman, F., Özhatayn N. & Koyuncu, M. (1983) *'New Allium Taxa from Turkey'*, *Notes RBG Edinburgh*, 41 (2)
Mathew, B. (1973) *Dwarf Bulbs*
——(1978) *The Larger Bulbs*
——(1986) *The Year-Round Bulb Garden*
——(1987) *The Smaller Bulbs*
Polunin, O. (1969) *Flowers of Europe*
——(1972) *Concise Flowers of Europe* condensed version of 1969 title
——(1980) *Flowers of Greece and the Balkans*
Polunin, O. & Smithies, B.E. (1973) *Flowers of South-West Europe*
Rix, E.M. (1983) *Growing Bulbs*
Stearn, W.T. (1944) *'Notes on the Genus Allium in the Old World'*, *Herbertia*, 11
——(1978) *'European species of Allium and allied genera of Alliaceae: a synonymic enumeration'*, *Annales Musei Goulandris*, 4
——(1981) *'Genus Allium in the Balkan Peninsula'*, *Bot Jahrb Syst*, 102
Stearn, W.T., assisted by Campbell, E. (1986) *'Allium'*, *European Garden Flora*, Vol. 1, 88
Synge, Patrick M. (1961) *Collins' Guide to Bulbs*

Wendelbo, P. (1963) 'Studies in Oriental
Liliiflorae', *Nytt Magasin for Botanikk*,
10

——(1967) *'The Genus *Allium*', *RHS
Lily Year Book*

——(1977) *'*Tulips and Irises of Iran*'

Alliums of Central Asia and the Far East

The literature of these regions tends to
overlap in part. Accounts of *Allium* in the
Far East are relatively infrequent. The
troubled political climate in the Far East
through most of the present century has
resulted, among other far-reaching reper-
cussions, in the decline of Western plant
collectors retrieving material and the sub-
sequent recording or reclassification of
endemic species.

Many 'new' plants from Japan seem to
be appearing in gardens. While the camera
can impart a degree of information,
English translations of some recent publications would be appreciated.

Similar translation is anticipated of the
Chinese flora. Westerners are beginning to
visit the Chinese mainland with increasing
freedom, many on horticultural tours.
Meanwhile the literature leans heavily on
plant material gathered around the turn of
the century and earlier, in the heyday of
Forrest, Wilson and company.

The works of Regel and Vvedensky have
already been covered in the previous section. N.M. Przewalski, 1939–1888, was
an eminent botanist who collected
throughout Central Asia and Mongolia.
His specimens were studied by Carl Maximowicz, 1827–1891, who had himself
travelled in the Far East, before eventually
publishing *Primitivae Florae Amurensis*.

'On the Alliums of India, China and
Japan' was published in 1874 in the
Journal of Botany, Vol. 12, by John
Gilbert Baker, 1834–1920, a botanist
working at Kew. Baker was also respons-
ible for individual accounts of *Allium*

appearing at this time in the *Botanical
Magazine*.

A Flora of British India by J.D. Hooker
appeared in 1892. This was revised and
supplemented by W.T. Stearn in *Herbertia*, 12 in 1945.

1944 had seen the publication of *Flor-
istic Regions of the USSR with reference to
the Genus Allium* and *Nomenclature
and Synonymy of Allium odorum and A
tuberosum* also by Stearn. This was later
followed by an important paper on '*Allium
& Milula* in the Central and Eastern Himalaya', 1960 in which several new species
were described, alongside photographic
records. Several maps illustrating distribution are included.

Stearn further produced the section on
Allium in *Enumeration of the Flowering
Plants of Nepal*, Vol. 1, eds. Hara and
Williams, 1979.

Professor Per Wendelbo wrote several
papers, listed below, covering areas of
Afghanistan and South-West Asia.

Meanwhile the Index has been pub-
lished in English of the *Flora Reipublicae
Popularis Sinicae*, Tomus 14, 1980, J.M.
Xu, in which 99 *Allium* are featured.

Finally the travels, prejudices and idiosyncrasy of Reginald Farrer should not be
forgotten. He was not overly enthusiastic
about *Allium* as a genus. One or two exceptions were made, however, and he is
also credited in conjunction with William
Purdom with the collection and introduction of certain species during their travels
in Gansu (Kansu) in 1914. Although for-
ever associated in the alpine gardener's
mind with the European Alps, he was to
die in a remote Burmese valley in 1920.
The English Rock Garden, 2 Vols., was
published in 1919.

A FURTHER SELECTION

Ekberg, L. (1969) 'Studies in the Genus
Allium II: A New Subgenus & New
Sections from Asia', *Bot Nosier*, 122

Ohwi, J. (1965) *Flora Japan*
Polunin, O. & Stainton, A. (1984)
*Flowers of the Himalaya
Wendelbo, P. (1958) 'Liliiflorae' in *Symbolae Afghanicae IV*, Biol. Skr. Dan. Vid. Selsk, 10
——(1966) 'New Taxa & Synonyms in *Allium* & *Nectaroscordum* of South-West Asia', *Acta Horti Gotoburgensis*, 28
——(1967) 'New Species of *Allium* from West Pakistan', *Nytt Magasin for Botanikk*, 14
——(1968) 'Some New Species of *Allium* (Liliaceae) from Afghanistan', *Studies in the Flora of Afghanistan*, 9, Bot Notiser, 121
——(1969) 'New Subgenera, Sections & Species of *Allium*', *Bot Notiser*, 122

Allium of America

It is hardly surprising that a continent as large as North America, with such diversity of climate and geography, does not have a universal flora. Although the American Rock Garden Society was founded around the same time as the Alpine Garden Society in Britain, the relative membership may reflect the distance one gardener may be from another. Enthusiasts in New England are a long, long way from their equivalent in California or Washington State. In Britain, members may travel from Scotland to London any weekend for a mini-conference, and frequently do. European plants are quite easily accessible, in the wild or in shows, and relevant literature proliferates. In North America plenty of small regional publications are published, with fewer potential readers for each one.

Hortus III, 1976, produced by the Bailey Hortorium Staff, is similar in format to the *RHS Dictionary* with compressed sketches of around 130 species, plus a listing of synonyms.

Sadly, while North America has so many delightful *Allium* to choose from, the genus is more neglected than one would anticipate. As if to compensate, the literature available has some most excellent *Allium* references.

The late Dr Marion Ownbey was responsible for so much of the material published. 'The Genus *Allium* in Arizona', 1947, 'The Genus *Allium* in Idaho', 1950 and 'The Genus *Allium* in Texas', 1950, published as Research Studies of the State College of Washington, while illustrated only by maps, are packed with information for the committed *Allium* collector.

A series of superb line drawings accompany the chapters (in Part 1, pp. 739–60) he wrote for *Vascular Plants of the Pacific North West*, C.L. Hitchcock, A. Cronquist, M. Ownbey and J.W. Thompson, 5 vols. to 1969. Twenty-nine species are fully described, with a key and illustrations; these pages set a standard for plant identification very hard to fault.

The same format is found in *Intermountain Flora*, A. Cronquist, A.H. Holmgren, N.J. Holmgren, J.L. Reveal, P.K. Holmgren, 1977, where Ownbey again supplied the information on 22 species for an area centred on Idaho.

Continuing down the Pacific Coast Ownbey's work is also found in *A California Flora*, ed. P.A. Munz, 1959 and *Supplement*, 1968, covering compactly 38 species, unfortunately without illustration.

The Illustrated Flora of the Pacific States, L. Abrams, Vol. 1, 1954, provides neat, clear line drawings and concise description for each of the 45 species covered. One or two of the species have suffered name change, though this is easily discernible from the synonymic entries.

Wildflowers of the United States, 1966, by H.W. Rickett. Six Volumes, consisting in total of 15 books, comprise this monumental work, Vols. 3–6 being richer in *Allium* reference. The numerous illustrations are colour photographs, the text simplified as a guide to identification.

The Peterson Field Guide Series—pocket-sized books packed with information and clear illustrations, both coloured and line—have provided two excellent volumes for *Allium* identification as well as other regional plants. *A Field Guide to Pacific States Wildflowers*, T.F. Niehaus, illustrated by C.L. Ripper, 1976, and its companion, *A Field Guide to South Western States and Texas Wild Flowers*, include a number of *Allium* species.

H.E. Moore in *Baileya*, 1954–5 covers *The Cultivated Alliums* available in the United States, providing a key and numerous illustrations. Botany, synonymity and the idiosyncrasy of each *Allium* Section is covered in a most interesting paper.

Finally, Mark McDonough provided a most readable series of 'Allium Notes' illustrated by line drawings for the *Bulletin of the American Rock Garden Society*, spread over four issues in 1984–5.

Numerous local flora cover the plants of specific regions.

Conservation

The conservation of plant and animal life that formed part of food chains was taken seriously by primitive peoples. Preserving a balance between immediate food requirements and future breeding stock was of obvious importance. Today all plant and animal life of not direct and obvious use to man is at risk.

Public opinion by and large no longer approves of the digging of plant material in the wild. Some countries have lists of protected species; others, for example in the USA, have wildlife parks where even picking any flower is punishable. Identification can be impossible in many species in these circumstances. Attitudes have greatly changed since specialist societies published articles describing the making of plant-collecting equipment for Alpine holidays.

Reprehensible collecting of whole colonies of wild plants has been laid at the door of certain commercial horticultural dealers. Hopefully this too will become unacceptable, an outcome aided by gardeners boycotting the purchase of plants known to have been dug on the sly. Collectors have not specifically threatened *Allium* and very many that include vegetable crops will continue to flourish.

Fortunately most alliums propagate well from seed, both for the continuation of wild stocks *in situ* but also for the pleasure and education of gardeners and botanists. So many genera produce prodigal amounts of seed one does not envisage the imminent banning of seed collecting. Obviously removal of the majority or all of the fruiting stems of any wild stand of plants will threaten future generation; in years to come the ethics of seed collection may also need to be considered.

5 *Cultivation*

All cooks have a recipe for fruit cake; so plantsmen have special mixes for sowing, for potting and for cuttings. The majority of ordinary gardeners trust to luck and use whatever is to hand. A reasonable compromise between the two usually works quite well.

In wet climates peat based composts and Perlite tend to get too wet outdoors, while being excellent used in greenhouses or tunnels. For the majority of sowings additional sharp sand or grit available from garden centres and builders' merchants is required for drainage, vermiculite can also be used. Most alliums will want sharp drainage and while a mix of half sand or grit compost will do nicely for germination, the mixture may then dry too quickly, falling away when re-potting thus damaging fine roots.

My preference is for a commercial loam compost with added fertiliser, in Great Britain a John Innes No. 2, or the local alternative. Two parts compost to one part sand/grit and a dash of bonemeal answers very well. Allium seed being of a size that handles easily, approximately 8–10 seeds are sown round the periphery of a 7–8 cm (3 ins) pot, then covered with a generous cm (⅓ ins) of grit.

While fresh seed is divided into two parts, one sown immediately and the other in early spring, in reality most sowing is done in January, February and March, when the seed exchange packages arrive. The majority of pots are placed outdoors on raised stands and left until germination. Our climate is very wet but 'frost-infrequent'. In unusual years of extra chill seed sowing is delayed.

Alliums known to produce foliage in autumn or mid-winter are sown as soon as seed is ripe or available and watered accordingly. Seed from hot dry climates is left dry until early spring. Attempts are made to reproduce the availability of water in the plant's natural habitat. The detailed analysis of geographical variants described in *Growing Bulbs* by Martyn Rix is very helpful here. Unfortunately the difficulty of identifying the needs of each specific *Allium* has led to the death of many small defenceless seedlings. The information on habitat, flowering period and hardiness in the species entries in Chapter 7 may be used as a guide.

The majority of seed germinates in spring and early summer and is then brought into the greenhouse for feeding, watering, and later providing summer rest for species that require this. The majority of alliums that flower in spring and early summer become dormant, spending the summer months outdoors with overhead cover or in the greenhouse, the pots often heeled in under the staging. This prevents the compost from drying out completely. While leaving summer dormant bulbs outside to face a drenching English June/July may annihilate a promising bulb collection, lessons may be learned from examining the apparently hard-baked soil in which species such as *A. falcifolium* grow in the Siskiyou Mountains of Oregon. Some 7–10 cm (3-4 ins) below a sun-drenched surface the soil can be quite moist, just where the roots penetrate, while the layers are dust dry where the bulb lies.

Feeding the seedling bulbs in their first and second seasons allows them to build

A. falcifolium in Oregon (O'Brien)

up stores for later replanting in individual pots before their third year. Bonemeal in the potting compost may be adequate for the first season for slow-growing small species. However, a gritty compost allows nutrients to leach out of the pot, so watering with a liquid feed such as Phostrogen may be necessary. Plants in active growth have an increased ability to utilize nutrients and it is advantageous to the small plant to build up tissue in its first year as reserve for the dormant period to come. Remembering that small human babies cannot digest a meal of prime steak, so the young plants should not be given a diet too strong for their wellbeing. Doubling the degree of extra feeding may not be beneficial and manufacturer's instructions should be heeded.

Some robust species that form slender bulbs or those on rhizomes may prefer replanting in the green as snowdrops do. Others that form globose solitary bulbs may be safer repotted while dormant but care must be taken to replant at the same depth in the new pot as in the old. Repotting toward the end of a dormant phase prevents the small bulb becoming too dry and shrivelled in loose, new compost. Watering from below by standing pots in shallow gravel trays or plunged in sand can avoid over-wetting the compost.

Winter watering may be difficult in a climate as changeable as that of the British Isles. Active growth may occur in mild spells midwinter, tempting the gardener to water pots, only to find that frost returns a day or two later. If pots are plunged in sand, grit or soil in greenhouse or frame and the pot content is sharply drained, little damage may be done. The depth of material around the plant slows down freezing of the water content of the medium, also that in plant tissues. Pots standing on staging without protection and filled with soggy soil may freeze in several hours, killing the occupant. Experience is required to gauge water requirements, overall safety lies in keeping pots on the dry side, watering by spoonfuls rather than with a lavish hand.

Consistently cold conditions require keeping pots nearly dry when in unheated frames or houses. Heat sufficient to prevent freezing, or glass insulated with polythene sheeting may be a good investment. Beginners and experts alike look forward to the day when really accurate weather forecasts become available. Growers in climates with heavy consistent snow cover can with greater safety presume less variable soil temperatures. Pots with no foliage above the surface are, as a general rule, safer left dry.

Pots sunk in the soil under greenhouse staging are comparatively easy to manage, requiring very little watering. The ground moisture is often quite adequate for plant requirements winter-through, provided the house does not sit on a soggy site. Earthworms can be prevented from gaining access through pot bases by rings cut out of zinc gauze placed over the hole. Slugs may be trapped in sunken containers of beer.

More mature bulbs also require watering consistent with their natural habitat. Plants from areas with autumn rain may come into leaf, remaining evergreen through winter, and will require water sufficient to support the foliage, the amount varying with the weather conditions of their new homes. Many may remain in outdoor beds without problems. Those from hot dry countries with rain falling mainly in spring or from snow melt may rot if allowed to become wet; for these, if requiring frame or pot cultures, winter watering should be minimal or withheld. As growth shows in spring watering can be plentiful, to be decreased as foliage withers and plants withdraw into dormancy. During this period pots should not be allowed to become completely dry, occasional trickles, preferably by basal watering, can be given, but always erring on the side of restraint.

Records over the years suggest that throwing out pots that have not germinated by their third year is justified. While many seeds germinate several weeks after sowing, others have missed the vital moment and start into growth the following season. I have not observed germination in any allium seed after a second year from sowing but other growers may disagree with this comment.

Seed is easy to gather, but may take time to separate from the chaff in many small species with packed umbels. Packaging in paper, not polythene, envelopes in airtight containers in the salad drawer of a domestic fridge has provided an adequate method of storage.

Species that form bulblets or bulbils are readily propagated by growing on, using the same conditions suitable for seedling bulbs. Rhizomatous and hardy border alliums can be divided in similar fashion to herbaceous perennials, the robust species being replanted in the required sites, smaller rock garden plants likewise or grown on in pots.

Propagation

When, following greenhouse care, a plant has flowered, been photographed, measured and hopefully identified (a very hit-and-miss affair, most baffling if the seed bears no resemblance to the identity assigned to it by the harvester), I usually plant it in the open garden to assess its hardiness and garden worthiness.

Under no circumstances however, should one plant out an allium bearing bulbils in the head or with numerous bulblets surrounding the base of the stem without being very sure what one is doing. These are the diagnostic signs that warn of plant thuggery. Alliums like *A. vineale* or *A. carinatum* may overrun a garden in a single season.

With the exception of *A. ursinum* and similar species, alliums prefer open

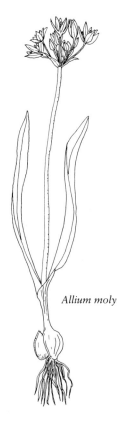

Allium moly

tumn, impervious to or even enjoying summer rainfall. Species that require a period of rest but are sufficiently hardy for life in the open garden may find the summer moisture absorbed by tree and shrub roots provides conditions to their liking.

Larger bulbs require periods of dormancy, perhaps from late spring but ordinarily in the summer months, after foliage and flowers wither. A number of the drumstick species flourish in well-drained sunny borders. As a rough guide, the larger the bulb—often those of plants inhabiting desert or semi-desert areas— the less moisture tolerated during dormancy, although many of the small alpine species with tiny bulbs also require water restriction. Pot culture, greenhouse or frame allows drying-off periods to be controlled.

Pots of small species can be plunged into the rock garden during flowering, then lifted as seed begins to form and placed under greenhouse staging. Bulb frames that can be opened for spring rain then closed for dormancy are ideal for many bulbs. Alliums may however form bulblets too enthusiastically for unrestrained planting; again, plunged pots can prevent such problems. Autumn watering as discussed for seed growing may be necessary. A frame requires a base of rubble for drainage, over-layered with a gritty loam compost. Above this in a layer of coarse sand or grit, the bulbs can be planted to prevent rot around the crowns. A further topping of gritty compost with a surface finish of yet more grit makes an ideal growing area, which can be as little as 30 cm (12 ins) deep to accommodate small bulbs and raised to any height convenient for the grower. The covering glass frame must be manoeuvrable for watering when appropriate and allow for air circulation.

An ordinary cold frame may be quite suitable for housing moderately hardy bulbs that require only mild monitoring of moisture requirements, or for winter protection.

ground. This may mean exposure to full sun, remembering that species with fibrous roots and others like *A. moly*, whose foliage can burn, may require some shade in warm gardens. Most species do not cope well with competition and may actively hate ground-covering. Only the weedy ones will fight and win the battle in a horticultural slum!

The bulbs give some indication of the cultivation required. Fibrous roots and slender bulbs suggest that garden culture will create little difficulty. While moist acid soils suit many of the fibrous rooted species, the majority are happy with a warm, well-drained limy soil that does not dry out completely. Persisting foliage through most of the year also indicates that plants will probably be quite happy with this treatment. Plants in these broad categories often flower in late summer or au-

Suitable small bulbs can be grown happily in an alpine house, the later-flowering species providing colour when many of the normal occupants have become less decorative. Likewise, many of the larger bulbous species may be planted as bedding in autumn, then lifted after flowering, and dried or heeled in spare ground for replanting once more the following season.

Suggested cultivation is indicated in the species entries in Chapter 7, and suitable plants for garden situations in Chapter 6, the latter considering several easily obtainable species.

Farmyard manure, however well matured, is best avoided in feeding most bulbs. Fertilizers high in nitrogen encourage leaf formation, excellent for grass but not recommended for most alliums. Potash and phosphate will encourage healthy growth and flower formation; a tomato or rose fertiliser gives good results. Growmore has slightly more nitrogen, recommended in the open garden rather than for culture. Few alliums have been known to turn their noses up at a handful of bonemeal on planting or indeed the same in a top dressing, which is my own preferred method. Like humans, they enjoy tasty feeding.

Pests and Diseases

As a genus *Allium* seems surprisingly free from attack by pests or disease. The main enemies are moisture at the wrong season of the year and poor drainage at any time. A few alliums can cope with heavy rainfall given reasonable soil but waterlogging will kill most bulbs and alliums are no exception.

Plants contending with adverse physical conditions are more at risk of contracting diseases. Good cultivation, appropriate feeding and attention to adequate watering will go a long way to providing countermeasures to the hazards waiting to destroy one's precious plants.

Fungicides and insecticides change by the year as predators, both plant and animal, develop resistance, or because the toxicity to the marauder proves to be dangerous also to the gardener. For this reasons specific chemicals have not been recommended. Aldrin, for example, has been withdrawn from amateur use. Marketing has been governed in the UK under the Control of Pesticide Regulations (1986) within the Food and Environmental Protection Act (1985) and even stricter supervision is envisaged through Control of Substances Hazardous to Health Regulations (1988). Public opinion is inclining towards the abhorrence of chemical pollution. The problems that have afflicted birds of prey through food chain contamination have made rural inhabitants most wary of pollutants. The fells above our Lake District garden are home to the only pair of Golden Eagles breeding in England, a cogent reason for asking if *any* chemical aid is really necessary. So many proprietary products are labelled 'harmful' to fish, a euphemism for lethal if allowed to leach into watercourses and eventually into the sea. Good cultivation should breed healthier plants, perhaps the only solution to disease that amateur gardeners should favour, despite marketing pressures.

The use of animal deterrents or insect parasites is becoming increasingly popular as a counter-attack. Local agricultural colleges or consultants may be willing to advise the most effective action against plant disease or depredation.

Slugs are always partial to young growth; newly emerging seedlings may succumb and flowering stems may be nibbled through. Some species, *AA. sikkimense mariei* and *beesianum* have suffered in our wet garden, but slug damage has not been a great problem overall.

In the greenhouse small plastic steep-sided pots filled with beer sited between plants provide lethal drinking holes. Slugs die, gloriously intoxicated, in a liquid

grave. A greenhouse full of emerging allium shoots and beer traps has a pleasant
wintry aura suggestive of waiting plough-
men's lunches.

Moles, being carnivorous, do not
damage plants by eating them. Unfortunately undermining may cause serious
losses as bulbs dislike air tunnels running
through emergent roots. They also dislike
the effect of falling through airspaces and
landing at uncongenial depths. The dry
beds around our house walls should be
ideal for bulbs. Unfortunately moles have
established a hereditary right of way
through to the springs rising from underly-
ing rock lower in the garden. This has led
to many losses. Planting bulbs in flower-
pots sunk in the ground is an effective
remedy, others have used bottomless tins
or buckets to minimise soil disturbance.
Latticed containers for water plants are
also helpful not only against determined
tunnellers but to prevent multiplying colonies of bulbs from wandering unduly.

Strychnine, traditionally the poison used
to kill moles, is available only to agricultural users in the UK. Professional mole
catchers are effective on a short term basis,
but the moles rapidly repropagate. Desperate measures might involve worm destruction with copper salts to starve the animals
out, forcing them to move to pastures new,
but this scorched earth policy would
damage the soil and be merely a short term
measure, not to be recommended.

Break-back traps are unpleasant, not
necessarily instantaneously fatal, tedious
to set and increasingly expensive to replace.

Rabbits have not been a problem with
growing alliums—while myxomatosis
controlled their numbers—but mice enjoy
Allium bulbs when they cannot have their
fill of tulip or crocus. Half a kilogram (1 lb)
of onion sets was skilfully removed in 48
hours from our Lakeland garden, an area
where alliums as a vegetable crop are
rarely grown. The fibrous rooted alliums
are untouched.

Drumsticks in the border (Alice Lauber)

Mice and voles have since been virtually
eliminated by the multitude of feral cats
that live on the fellsides, encouraged
doubtless by the milk and scraps put out
for them, the bird food having been placed
well out of harm's way. A good mouser
was beyond value in the granaries of
Ancient Egypt. While no longer officially
worshipped, no bulb frame should be
without one.

In the USA, gophers may cause damage
both by burrowing and by eating bulbs.
Most countries will doubtless be plagued
by some form of energetic, hungry rodent
intent on finding a tasty well-flavoured
bulb.

Aphids infest many of the greenhouse
species and seedlings brought under cover.
While no great damage appears to be done,
potential infection with viruses should be
considered and if other susceptible genera
are grown then control is imperative. Sys-
temic insecticides also kill friendly insect
predators and squashing with fingers is
messy. As aphids will always be with us,
washing-up liquid may be the war weapon
of choice, the bubbles blocking the creatures' breathing apparatus.

Some pests are difficult to control in
areas where many plants are grown in
proximity. It is not appropriate here to
discuss methods employed by agriculture
though public interest is forcing a debate
on chemical usage.

Gardeners who still wish to use controlling agents may care to differentiate between the organic pesticides such as Natural Pyrethrum, Derris and Insecticidal Soaps or the synthetic pyrethroid Permethrin, all of which have minimal damaging effects but are effective against many pests including aphids, thrips and red spider mite, and the many non-organic products that flood the shelves of garden centres.

Red spider mite has not—to date—proved a problem with growing *Allium* in our cool climate and I have found neither Cutworms nor Vine Weevils in the pots of casualties, but this may be simply good luck.

Vine and Clay-Coloured Weevils, however, attack a multitude of plants, leaving nibbled circles along leaf edges. The adults feed on shrubs at night and an easy method of trapping is effected by surrounding the plant base by newspaper or polythene sheeting then shaking the branches. The creatures fall onto the layers, are easily seen by torchlight and may be desroyed. The larvae do more damage by eating roots and are lethal in pots and bulb frames. Root destruction causes rapid wilting and death of plants. Infestation should be suspected in any sudden flagging of foliage. The grubs are curved, white chubby objects with brown heads, often in groups. Soil should be sterilised if the means to steam heat are available, a course not easily followed by amateur gardeners. More practically, contaminated compost should be thrown out or burnt. Alpine houses are quickly infested, Primulas in particular being susceptible. Appropriate insecticides have been incorporated in potting compost, Aldrin which was effective has now been withdrawn from amateur use.

Unlike the drab Vine or Clay-Coloured Weevils, Lily Beetle is a beautiful sealing wax red, but just as damaging.

Delia antiqua, the Onion Fly, a serious pest in commercial onion crops, is more prevalent in warmer counties. Eggs are laid in the necks of the bulbs or emerging foliage. Emerging maggots eat into the bulbs, which become hollowed and slushy. Infestation becomes apparent in spring and early summer, when the foliage turns yellow before dying off. Two or three broods may be produced, the maggots taking approximately 21 days to mature and pupate. The chestnut brown pupae remain in the soil until hatching. Infected bulbs should be destroyed and the area sterilised. Ornamental species can be attacked, particularly in areas where Onion growing is prevalent and insecticides may be necessary.

Thrips cause mottling of foliage and distortion. Damage is commoner in warm areas and under greenhouse or bulb frame cultivation. Control may require insecticides.

One of the numerous Eelworms is parasitic on onions, causing the bulb to swell and go soft. Infected plants should be burned, and no allium should be grown on the same ground for five years. Chickweed is a host to the same pest, so onion beds should be kept weed-free.

Several fungal infections of vegetable crops also attack ornamental species. Fungicide dusts and systemics combat attack. Smut causes dark streaks suggesting blisters on both bulbs and foliage, the causative agent is *Urocystis cepulae.*

Botrytis species attack many garden plants. Infection begins in dead flowers, damaged or withered leaves and bulbs. Good garden hygiene in bulb frames or greenhouse is a prophylactic. Once *Botrytis* spores have formed emerging seedlings are at risk. Prevention and cure has often involved the use of fungicides. Benomyl (Benlate) has been widely used but in many areas fungal strains have developed resistance as they have to carbendazim and thiophanate-methyl. Sulphur dusts and copper compounds are old-fashioned remedies.

Another fungus, *Sclerotium cepivorum,*

causes White Rot: bulbs are covered with a whitish, fluffy film, as the foliage turns yellow.

Peronospora destructor, Onion Mildew or Downy Mildew shrivels the leaf tips. So many species have leaves that wither naturally during the flowering season, however, that fungal attack may be over-diagnosed!

Rust fungus was prevalent in 1988 on *A. crenulatum* in the Olympic Mts of Washington State.

Gardeners frequently refer to virus disease as if one omnipotent malefactor attacked plants. Reality is rather different, for as many different viruses exist as there are types of bacteria. Indeed their numbers may well exceed those of allium species! It is therefore quite inaccurate to refer to virus diseases in the singular. Even with modern laboratory facilities, identification of specific infection may not be possible. Treatment of plants may be impossible and only the destruction of diseased tissue will prevent spread. Lilies are martyrs to virus diseases, alliums fortunately seem fairly non-susceptible.

Several viruses are spread by aphids using their mouth parts as a syringe transmitting infection. Disease-free plants may be produced by micropropagation or by seed, few genera transmitting disease by the latter. Control of aphids is therefore important not just to the carrier but to other plants more at risk.

In conclusion, while pests and diseases will be always with us, plants, like humans, may benefit from acquiring immunity from infection. Good garden hygiene, adequate feeding, drainage and sunshine are of greater importance to plant health than any programme of 'killer medication'. Disliking all spraying and dustings, possibly an unexpected attitude in a medical practitioner, I find it salutary to remember that many human scourges, cholera, typhoid, plague were beaten by *sanitary* measures. Remembering too disasters brought about by alien species introduction, rabbits to Australia, feral cats and rats to islands with flightless birds, some caution may be wise before introducing predators such as nematodes to exterminate plant enemies, even such villains as Vine Weevils. Finally it is important to realise that all sprays and aerosols, whatever the contents, in home, garden or industry introduce small particles into the atmosphere and thence through the nose into the lungs. Whatever the spray, be it furniture polish, deodorants, fungicides or fly killers, clouds of inhalants are unlikely to be beneficial to the human operator.

For those interested in fuller details of insecticides and fungicides for the amateur gardener, several leaflets are published by horticultural societies. Technical data soon becomes outmoded, but several manufacturers publish leaflets on their products and agricultural colleges may be able to provide advice. A selection of books might include, Buczaski, S. and Harris, K. (1981) *Collins Guide to the Pests, Diseases and Disorders of Garden Plants*, and Ministry of Agriculture, HPDI *Diseases of Bulbs*, ADAS, (1979).

6 Alliums in the Garden

Within the enormous range of *Allium* species, plants can be chosen for most garden habitats.

Easy Alliums

Several species find their way into the cheaper bulb catalogues; generally they are a good buy and so cheap as to make seed sowing a waste of time.

For the rock garden, *A. oreophilum* var. *ostrowskianum*, usually sold as *A. ostrowskianum*, is a 10 cm (4 ins) tall charmer in the brightest of carmine pinks—not magenta. Plant in a dry, sunny spot. June.

A. sphaerocephalon, 60 cm (24 ins), July/August, reddish-purple, flattened, globular heads on thin, whippy stems are easy to grow in a sunny border.

A. aflatunense Hort, usually 75 cm (30 ins), May/June, is one of the easiest 'drumstick' or 'cricket ball' alliums with rosy-violet heads. E.A. Bowles used the latter term, but as cricket balls are rarely seen on stalks, the furry-topped, twirling stick of a drum major conjures up a more evocative image. Try the bulbs in a sunny border with good drainage, even in north-west England.

A. moly, buttercup yellow, 30 cm (12 ins), is often reported as invasive; to date I know no one who has found it so. Some clones are shy of flowering, pick the right stockist and look for *A. moly* 'Jeannine', a great improvement on the norm. The greyish-green leaves are quite attractive, disappearing in late summer. Grow-

ing well at the base of beech hedges in our wet garden, useful for naturalising between shrubs, it also flourishes in the rich soil of an allotment.

A. cyathophorum var. *farreri*, alias *A. tibeticum*, *A.* sp. Tibet, violet, nodding heads, 30 cm (12 ins), appears unperturbable and ubiquitous. Foliage is good-looking at flowering time; dead head if seedlings would be unwelcome. Useful on rock garden, peat or sunny patch with good deep soil; plant so that the roots can find the moisture they like. July.

A. flavum and *A. carinatum* subsp. *pulchellum* are lookalikes with long, unequal 2-valved spathes, varying in height from 10 to 45 cm (4–18 ins), the first yellow, the second rose-purple or white. The bell-shaped flowers on unequal stalks, the outer flowers drooping, produce a lovely fountain or firework effect. Suitable for sunny, borders not too dry. July, August, September.

A. cernuum, a nodding onion from North America, is almost evergreen, 45 cm (18 ins), June/July. The bell-shaped flowers vary from deep rose-purple, pink to blush. Sunny borders with some moisture.

A. schoenoprasum, Chives, available in several sizes, the best have larger umbels of a good colour. Two crops are produced almost every year and often three; the dead, chaffy flowering stems pull out, keeping the clumps tidy. A lovely form with deep clover-pink heads is called *A. s.* 'Forescate', which produces seedlings with violet heads. The cut flowers do not smell, while the leaves are excellent in soup, salads or sandwiches. Such excellent front-of-border plants can survive a heavy soil.

Alliums for the Rock Garden

The microcosm of a rock garden may mimic the full range of Earth's ecology.

Those gardeners who live with hot, dry summers and cold, dry winters, with consistent snow cover, will grow in their rock gardens plants suited only to the alpine house of others. Mediterranean climates cosset a range of plants endemic to Greece, coastal California or Asia Minor, plants that also find their way into the bulb frames of British gardeners.

Consider, however, the rock garden or raised beds of British growers able to provide the sharp drainage that counteracts high rainfall. *A. insubricum* (syn. *A. narcissiflorum* Hort.), from northern Italy, has ravishing deep rose flowers which hang their heads in June/July. The plant with its surprisingly white, naked bulbs is a tantaliser. Life on a raised gritty bed in Cumbria, with high rainfall, apparently pleases more than pot culture. Had I but one *Allium* to grow, *A. insubricum* would be it.

In the fairy-tale, Rumpelstiltskin challenged everyone to learn his true name; *A. insubricum* does likewise. Hanging its head, it remains *insubricum*. Were it to lift its head after seed forms, it would be *narcissiflorum*. The remaining story rests, as in a fairy story, in a key, but alas this key is botanical.

A degree of magic lingers with *A. olympicum*, despite some taxonomic problems. The Greek Pantheon demands an *allium* ten foot (3 m) tall. *A. olympicum* reclines on stony ground. Grey-green leaves hug the gravel to produce in June heads of clear pink flowers, hardly raised above their gravelly bed. Long stamens enhance one of the neatest onions to grace a rock garden.

(Top left:) *A. oreophilum* var. *ostrowskianum* in Colthurst garden, Waddington, Lancashire (Author) (Bottom left:) *A. schoenoprasum* 'Forescate' in Preston, Lancashire (Author)

Growing *allium* from seed is one of life's lotteries, *A. olympicum* confirmed a lifetime faith in serendipity. *A.* sp. JCA 361 is similar, painted in softer colours. Both these small alliums grow apace in Cumbria on a raised bed, slowly increasing but rarely producing seed in the wet climate.

Damp corners of the rock garden are perfect homes for the small Far Eastern species—*AA. amabile, mairei, sikkimense, beesianum* and *cyathophorum* var. *farreri.*

A. karataviense forms a splendid border to a gravel path in the rock garden in Edinburgh's Royal Botanic Garden. Rather sharper drainage is required in north-west England; plants fare better with protection.

Alliums as Pot Plants

Several alliums have foliage to grace a pot. The group includes *AA. karataviense, mirum, protensum, schubertii* and *bucharicum*, with broad, firm, green or glaucous leaves, presentable at flowering time plus pleasing flower heads. Leaves appear in late winter or early spring then April finds the plants in flower. Finally the heads in seed are beautifully structured open spheres, retaining their shape and texture, often for years. Bulbs of the rarer plants are beginning to appear in specialist growers' lists.

A. christophii also fits a pot, the stem being rather longer than those of the previous group. Bulbs are cheap and freely available. The dried heads are excellent for flower arranging.

Plants for spring flowering would include *A. paradoxum* var. *normale*, a crisp white with large flowers. *AA. hyalinum, pendulinum, zebdanense* and *triquetrum* have open sprays of white stars.

Some of the American alliums make superb pot plants; *A. unifolium* with sugar pink tepals on stout stems, *AA. acuminatum, peninsulare, falcifolium* and *dichlamydeum* are all most attractive and

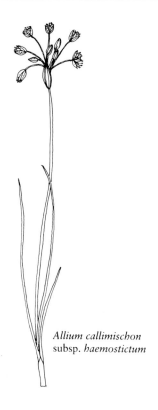

Allium callimischon
subsp. *haemostictum*

Several of the forms of *AA. flavum, paniculatum* and *pallens* are small enough to be suitable for pot growth. *A. pallens* may be doubtfully hardy in the garden and will positively benefit from protection.

A. callimischon, and *A. virgunculae* flower late in autumn, while *A. chamaemoly* heralds the turn of the year, one of the early flowers of pre-spring. Pot growth allows the grower to savour the small dainty plants.

Alliums in the Bulb Frame

Controlled weather conditions suit *Allium* very well, since moisture in the growing season and restricted water during dormancy are easily attained. Care is needed that species that reproduce rapidly are not introduced unknowingly into a bed and are limited by pot culture. Seeding should also be strictly monitored.

The American species listed for pot culture, plus most of the high altitude and desert species, are satisfactorily grown this way.

A frame provides protection for small, late-flowering species which while reasonably hardy would be muddied by autumn wind and rain. Their water requirements differ from those of the spring-flowering bulbs.

Dry summer rest is essential for most of the Middle Eastern and Central Asian species, including *AA. akaka, bucharicum, cardiostemon, mirum* and *schubertii*, all of which are attractive plants for frame or alpine house.

Some of the larger species are difficult to accommodate in pots or frames. Dormancy for these can often be achieved by growing plants in the lee of deciduous shrubs, or in bedding around houses where the eaves of the roof prevent rain soaking the ground.

colourful. After seed production, the bulbs will require restricted watering and a summer rest.

Many of the small alliums grown from seed profit from one or two seasons' pot-life while building strength and numbers. Eye-catching blue Asiatic *AA. sikkimense, beesianum* and *cyaneum* are all amenable, requiring adequate watering for they do not like the compost too dry.

A. mairei and A. amabile need the same conditions, their flowers are shades of white and pink. (All these Asiatics enjoy garden conditions with moist drainage and possibly shade.)

Summer bulbs from the rock garden might include *A. olympicum*, pale pink and ground-hugging. *A. insubricum*, often obtained under the erroneous name of *A. narcissiflorum*, is a beautiful allium for midsummer.

Alliums for the Border

The weird heads of *A. ampeloprasum* var. *babingtonii* were favoured plants of E.A. Bowles for the back of herbaceous borders. This species is fully perennial. Not having found the bulbils detach readily from the main umbel, I cut the stems when autumn rains bedraggle any remaining flowers. Dried heads then spend the winter with other deads, the secondary bulbil flowers keeping their pink tints for many weeks.

Flowering plants of the Common Leek are also perennial if left undisturbed. Every summer 120–150 cm (4–5 ft) stems have added height and weight to the rear ranks of herbaceous plants. These heads too will dry satisfactorily for a single season, after this they look a little shabby.

Few people are aware that garlic has a flower. The circular heads are decorative when the beaked spathe has recently released the flower head. Then it sits atop the umbel looking for all the world like a Dickensian nightcap. Whatever the cloves may do in the kitchen to scent the air, in the garden garlic is as well behaved as any other flower. The bulbs require a warm, dry corner to thrive and behave as perennials. The climate of Victoria, Vancouver Island suits them well. Away from the Mediterranean lands, bulbs may rot in winter wet and cold unless perhaps treated as a bedding plant. The Garlic Farm offers a red skinned clone of garlic which is claimed to be especially suited to the British climate and resistant to rot. Pink or greenish-white, occasionally the flowers are deeper hued. July, 60–70 cm (24-28 ins). In some forms the heads may contain a surfeit of bulbils.

The soft yellow of *A. obliquum* is a valuable link between vivid colours in the border. A middle ranker, it flowers mid-summer reaching 60–75 cm (24–30 ins).

Several of the large 'drumsticks' will flower well in a sunny, well-drained border. The tallest is *A. giganteum*, growing to 180 cm (6 ft) in ideal conditions.

Hardier 'drumsticks' are easily obtainable from popular bulb stores. The allium sold as *A. aflatunense* should flourish in most open borders; bulbs labelled *A. rosenbachianum* and the easier *A. stipitatum* may include other species. *A.* 'Purple Sensation' is an excellent garden plant. For most gardeners, trial and error will indicate which 'drumsticks' will grow well in each particular site, without undue concern as to correct identification. Some white forms are obtainable but the majority are violet-purple.

A. sphaerocephalon has already been mentioned, *AA. polyastrum, macranthum, tuberosum* and *wallichii* will all enjoy the middle ranks of the border when not kept too dry.

A. caeruleum (syn. *A. azureum*) and *A. caesium* in differing intensities of blue, 15–20 cm (6–10 ins) tall, will suit a dry, sunny

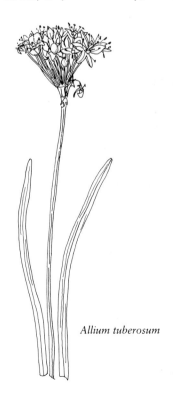

Allium tuberosum

border, echoing the blue and springy form of *Catananche*. Some sunshine but an average open border that does not dry out too quickly will suit *AA. flavum, carinatum* subsp. *pulchellum* and *paniculatum* in shades of yellow, purple, white and pinky-beige.

The football heads of *A. christophii*, starbursts of brittle fairground fantasies line flagged paths with splinters of purple, drying to improbable airy spheres. The umbels can measure 30 cm (12 ins) diameter, superlative winter dried decoratives. Flowering in May/June, in northern gardens the bulbs may be lifted when entering summer dormancy, dried and replanted in autumn as bedding tulips are.

As stocks become available the stunning heads of *A.* 'Globemaster' will create an even more spectacular display.

Alliums for Damp Sites

Among lists of bog plants, alliums are conspicuously absent. In the family of Liliaceae, to which *Allium* belonged until recently, there are several genera from which species can be chosen to inhabit damp locations—*Hosta, Smilacina, Disporum, Polygonatum, Zigadenus, Camassia* and some lilies. Few *Allium* are quite so obliging.

A. validum, the Swamp Onion, grows in alpine and subalpine meadows or swamps, but at a height of 1,700–3,700 m (5,000–11,000 ft), from California to Nevada, Washington State to Idaho. During the winter at these heights land and plants are snow covered, ensuring winter dryness. Springtime moisture for plant resurgence comes from the snow melt. So many alpine plants utilise this running water, flower, attract fertilising insects, then concentrate on seed production in the weeks when the high ground is draining, often baking in the sun. Later in summer plants die down, leaving only leaf tips above ground. If these conditions can be duplicated in the garden

the range of plants grown is immense. Unfortunately winter in many gardens is alternating damp and frost without snow cover. In these circumstances drainage is important, free-standing moisture or bog will bring a death sentence to plants who enjoy water during their season of active growth.

The recipe therefore for *A. validum* and the similar *A. brevistylum* is a damp area that has adequate drainage in winter but moisture in the summer. In dry gardens summer moisture must be provided. Both these plants are about 45–60 cm (18–24 ins), with thick, iris-like rhizomes. *AA. goodingii, eurotophilum* and *plummerae* are similar species and may settle happily in a dampish spot.

Many of the smaller alpine species require moisture in early spring but sharp drainage in winter and a summer rest. As a general rule, those groups of alliums with slender bulbs clustered on rhizomes are tolerant of moisture to a far greater degree than those with discrete bulbs. Alliums with large bulbs are in the main adapted to conditions in which water is scarce for all or part of the year. Many alliums from the Far East are accustomed to a wetter climate, growing well in north-west England.

In a garden with acid or clay soil and an average rainfall of up to 300 cm (120 ins) alliums that must have moisture for successful cultivation require a degree of drainage unnecessary in drier climates.

Small, Far Eastern species that enjoy moist ground include *AA. sikkimense, cyaneum* and *beesianum*, all of which are a good clear blue, and small enough for the rock garden. In sunny, dry gardens a little shade may be well tolerated. *A. mairei*, and the possibly synonymic *A. amabile*, 10–25 cm (4–10 ins) with pale and deep pink flowers, enjoy the same conditions. Several of the Asiatic species flower after midsummer increasing their value in the rock garden. *A. cyathophorum* var. *farreri* is violet-purple, a little taller and an easy plant for acid soils or damp areas, quite prepared to flower in shade. This allium is frequently sold as *A. farreri* or *A.* sp Tibet.

Growing well in damp borders or in clay soils are *A. macranthum* rose-purple, 45 cm (18 ins), *A. polyastrum* (p. 162), 60–90 cm (24–36 ins) and *A. wallichii*, 45 cm (18 ins), both violet-purple, and white *A. tuberosum*, 50 cm (20 ins), which is equally at home in the herb garden.

Easy, good-natured *A. angulosum*, 45 cm (18 ins), mauve with evergreen leaves enjoys a moist or heavy soil, flowering well in a little shade. Many plants sold as *A. pyrenaicum* are *A. angulosum*. *A. schoenoprasum*, particularly var. *sibiricum*, thrives in similar conditions though sunshine encourages heavier flowering.

Several species allied to *A. fistulosum*, the Welsh Onion or Japanese Bunching Onion, grow well in acid, damp soils, flowering with whitish or yellowish umbels, around 45–60 cm (18–24 ins). Certainly not in the top flight of ornamental border plants, at least they can be eaten if failing to please. Equally at home in damp corners of the herb or vegetable patch are *AA. altaicum*, *obliquum* and *galanthum*.

Leeks have always been popular vegetables in northern England. Flourishing in clay, *A. ampeloprasum* var. *babingtonii* creates a talking point, rearing from the border with its manic head of flowering bulbils. Wild plants are also found in ditches close to the Atlantic seaboard of Cornwall and West Ireland.

For damp woodland *AA. ursinum*, *tricoccum*, *triquetrum* and *pendulinum* do well. Perhaps far too well in the case of *A. ursinum* and *A. tricoccum*, Ramsons and Ramps respectively; both are able to make the air hum with the smell of garlic in spring. Neither should be introduced into the civilised areas of the garden unless a takeover is acceptable. *A. triquetrum* requires slightly drier conditions and is not so rampageous in the North.

An unrelated selection of alliums will flourish in clay soils and appear to be happy in areas of high rainfall, provided drainage is adequate. Included are *AA. flavum*, *carinatum* ssp. *pulchellum*, *paniculatum*, *cernuum* and *lineare*. *A. carinatum* revels in soggy wet, heavy soil in the shade, nothing diminishes its will to live and the plants cannot be recommended. Neither is *A. paradoxum*, with its head full of bulbils. *A. paradoxum* var. *normale* is quite another matter. This is a lovely plant with large, pendent flowers reminiscent of Lily-of-the-Valley, flowering in spring attended by its bright green leaves.

Alliums to Avoid

Any *Allium* with bulbils in the head should be viewed with suspicion. These tokens of fecundity seem designed to convert the

Allium vineale, A. carinatum, A. scorodoprasum

nimblest fingers into thumbs, before rolling away with glee to produce the next generation. Principal villains, because of their ubiquity, are A. *carinatum*, Keeled Garlic, scourge of heavy soils, A. *vineale* and A. *scorodoprasum*. Identify them by the wispiness and number of their spathe valves, and the colour of their progeny. Next incarcerate them in heavy duty plastic or tins, dump into the incinerator or bin, grind them underfoot in the centre of a large concrete carpark, then wait for the inevitable germination next year.

Lesser thugs might include bulbilliferous forms of AA. *paradoxum, roseum, canadense, ampeloprasum* var. *bulbiferum*. Bulbils may also occur in forms of AA. *nigrum, caeruleum* or *moly*. This list does not exhaust the possibilities.

Gardens with a Mediterranean climate, albeit Naples or California, may find A. *neapolitanum*, A. *triquetrum* and others may romp away if the temperature is right.

Despite these warnings, many *Allium* growers have no problems with unwanted bulbils. Like humanity, for each antisocial type there are hundreds of well-behaved individuals.

Colour Ranges

As a genus, *Allium* appears in many colour guises. Through the spectrum, orange is unrepresented and shades of red and indigo are rare. Green appears in spathes and opening buds or as striping against other shades.

Red as a colour is restricted to comparatively few genera; tinges are seen in A. *glandulosum* found on the borders of New Mexico, the geographic area of *Phlox mesoleuca*. Colouring in plants has been speculatively linked with the preferences of pollinating insects. Are deep shades against the dun matt wastes of desert, highly effective insect lures?

Yellow colours A. *flavum* and

A. *pseudoflavum* from Europe, A. *moly* and softer-hued A. *obliquum*. Uniquely in North America is A. *coryi* from Texas. AA. *horvatii, hymettum, scorzonerifolium, chrysanthum, chrysonemum, chrysantherum, pervestitum, calocephalum* and *luteolum* fall in this colour range but many yellows are either not in cultivation or difficult to obtain. Pale creams to offset hotter colouring can be found in A. *fistulosum* and A. *ericetorum*.

Some forms of A. *dictyoprasum* (syn. A. *viride*) found from the Caucasus to the Middle East may have greenish segments with white edges. A. *paniculatum* can present in many colour combinations, beiges with green or brown, hints of pink and copper, the flowers repaying close scrutiny at eye level.

The few blue alliums are prized plants for the garden. Midsummer brings A. *caesium* and A. *caeruleum* for sunny, well-drained borders. Other blues originating in the Far East, in wetter regions, are remarkably easy for damp northern gardens. Flowering in May, A. *sikkimense* (syns. A. *kansuense* or A. *tibeticum*) is followed in August by the small A. *cyaneum*, with A. *beesianum* brightening September.

One of the deeper violet-tinted flowers, A. *cyathophorum* var. *farreri* is easily grown. Rather taller are A. *polyastrum* and A. *wallichii*, while A. *macranthum* has a grape-like bloom on its flowers. There are pale lilacs, mauves and rosy-pinks in abundance; A. *pulchellum* is a rich wine-purple, A. *cernuum* shades through blush to deep raspberry. The softer glow of A. *insubricum* is quite ravishing.

From the many whites the tepals of A. *paradoxum* var. *normale* have the crisp sparkle and texture of Poet's Eye Narcissus. A. *neapolitanum* is sold in Italy as a cut flower, its dainty flowers on slender pedicels a clear white. A. *pendulinum* and

A. *pendulinum* (Author)

A. triquetrum display a cool green stripe. *A. callimischon* ssp. *haemostictum* sports rusty speckles on a pale ground, in autumn.

Alliums as Cut Flowers

Not all alliums carry their onion trademark to the nose. AA. *subhirsutum, caesium, zebdanense, tuberosum* and *suaveolens* are softly fragrant. *A. neapolitanum* has a very faint scent, pleasant to those who like freesias or astrantias. Many species are quite without any odour and can be picked for flower arrangements for the house. Humble chives from the vegetable garden are often found with good flower heads, mixing well with summer arrangements. *A. schoenoprasum* var. *sibiricum* and the variant known as *A.* 'Forescate' are quite lovely. Taken to a hospital sufferer, all they collected were compliments; their identity undetected no one threatened to eat them.

Fingers may smell slightly after picking but the odour does not persist on the cut flower. A few advertise more blatantly. Avoid *A. tricoccum* and *A. ursinum*. *A. ampeloprasum* var. *babingtonii* carries a hint of Garlic that I admit to enjoying. A few heads are unoppressive, though an armful would be something else again.

The heads of *A. karataviense* and *A. christophii* are excellent for dried flower arrangements—lovely spiky spheres. *A. peninsulare* and *A. acuminatum* retain plum-coloured staining of the papery dried tepals, year after year. *A. unifolium*, bleached as bones, holds its flat wide petals on open umbels. Rounded, chunky *A. splendens* and *A. lineare* persist as solid knobs of parchment for small designs. The strange heads of *A. ampeloprasum* var. *babingtonii*, filled with solid little bulbils, are striking in a large composition. Most of the 'drumsticks' hold their stiff globular outlines, retaining black glistening seeds for contrast. These frequently have to be picked out for harvesting, suggesting that natural propagation follows detached heads rolling across dry ground to germinate far from the parent. *A. schoenoprasum* and others drop their seed lightly from the heads forming multiple offspring around the parent clumps.

The heads of both onions and leeks if left to bolt, from intention or negligence, are excellent cut material. They dry quite satisfactorily, slowly turning a pale shade of parchment.

When gathering for dried material, like many other genera alliums retain colour and stem consistency better if picked before ripe and in dry weather. Many will last for several years, requiring to be thrown away only when accumulated spiders' webs spoil the overall effect.

7 A–Z of Selected Allium *Species*

The description of species following includes bibliography based primarily on *Flora Europaea* (*Fl Eur*), *European Garden Flora* (*Eur Garden Fl*), *Flora Iranica* (*Fl Iranica*), *Flora of Turkey* (*Fl Turkey*), *Flora USSR* (*Fl USSR*), *Vascular Plants of the Pacific North West* (*Vasc Pl Pac N W*) *Illustrated Flora of the Pacific States*, *Wildflowers of the United States* (*Wildflowers US*), *Botanical Magazine* (*Bot Mag*) and *Flora Reipublicae Popularis Sinicae* (*Fl Rei Pop Sin*). The abbreviations used are in parentheses; the numbers used are species' entries not page numbers. The abbreviation HZ indicates hardiness zone, see p. 155.

Meanings of some of the species names have been included. Commemorative endings may include *i* or *ii*, *ianum*, *iana*, *anum*, *ae*, *is*; in these species or when the derivation is obvious no explanation is included. The occasional species' name has eluded detection.

While physical description has been kept as simple as possible, details of stamens, seeds and chromosome counts have been omitted. These can usually be obtained from the bibliography following each entry. Classification into Sections has not been included either as not all authorities are unanimous.

A. aaseae
Ownbey 1950

This small onion is on the endangered species list for Idaho, the only State where it has been found. As alliums propagate so well from seed, *A. aaseae* is included for hopefully plants will appear in cultivation before too long.

The stem is usually less than 5 cm (2 ins) tall, topped with deep pink flowers which fade to white, then later turn papery as they form seed. The leaves are longer, still green at flowering but withering as the seeds ripen.

Found on gravelly river banks on low hills around Boise in Idaho, flowering in April, this small onion would make a desirable plant for troughs. HZ 5

Named after Ownbey's co-worker on *Allium*, Dr Hannah C. Aase.

A. simillimum is a closely allied species, even smaller, with paler flowers and blooming later in the season.

SOURCES Ownbey, 'Genus *Allium* in Idaho'; Davis, *Flora of Idaho*. Illus: Hitchcock *et al.*, *Vasc Pl Pac N W*; Cronquist *et al.*, *Intermountain Flora*.

A acuminatum
Hooker's Onion, Tapertip Onion
Hooker 1839

This small bulbous plant was probably the onion mentioned in the entry for 8 June 1792 in *Menzies' Journal of Vancouver's Voyage*. Archibald Menzies was the Scots surgeon-naturalist on the expeditions to the Pacific coast of North America.

Flowering May to July, among dry, sunny rocks in North-West America from British Columbia to northern California, Montana, Colorado to Arizona, the purple-rose, many-flowered, open umbels are quite showy, though the leaves have usually withered by flowering. The sharply pointed tepals, the outer recurving, make for easy identification. 10–30 cm (4–12 ins).

A. *acuminatum* on Table Mountain, Ellensburg, Washington State (Author)

The flowers dry well; even after six years the papery tepals retain traces of deep pink staining. A lovely, rare form grew in the Boulder, Colorado garden of the late Paul Maslin. A crisp, clear white, this would be a most desirable plant for the rock garden.

Plants have not proved hardy in a wet, cool climate; bulbs tend to rot in high rainfall, despite sharp drainage. This is not surprising since in their native desert conditions bulbs are usually found growing 5–7 cm (2–3 ins) below the surface. Cultivation is recommended in frame or alpine house with summer rest. HZ 2

SYNONYM   *A. cuspidatum.*

SOURCES   *Hortus III.* Illus: Clark, *Wildflowers of the Pacific Northwest;* Hitchcock *et al., Vasc Pl Pac N W;* Cronquist *et al., Intermountain Flora,* 12.

A. acutiflorum
Loiseleur-Deslongchamps   1809

Ovoid bulbs send up a 15–50 cm (6–20 ins) stem, with 2–3 linear, flat leaves clothing the lower stem. The hemispherical umbel, 1.5–5 cm (³/₅–2 ins) in diameter, is packed with purplish, bell-shaped flowers. As the name suggests the tepals are acutely pointed. Yellowish anthers top stamens that are shorter than the tepals.

Plants grow on rocky or sandy ground on sea coasts mainly, from southern France, Corsica, through to north-west Italy. June–August.

George Don in his Monograph of 1827 observed: 'This beautiful species I have seen no where but in the Chelsea Botanic Garden.'

Cultivation requires dry, sunny, well-drained sites. HZ 3

SOURCES   *Fl Eur,* 85; Grey-Wilson & Mathew, *Bulbs.*

A. aflatunense
B. Fedtschenko   1904

A degree of confusion surrounds the identity of the tall 'drumstick' alliums. Introduction of the plant widely known as *A. aflatunense* is credited to the firm of Van Tubergen. Bulbs are easily available from most of the popular catalogues and prove trouble free, certainly in warm southern gardens, though in the north planting requires good drainage.

While the vital statistics of *A. aflatunense* are listed in *Flora USSR,* many if not all of the bulbs in commerce may be impostors. Probably few practical gardeners will be disturbed by this, content to grow an attractive, relatively troublefree, decorative Onion.

Under the house flag of '*A. aflatunense* of commerce' may well be found *AA. stipitatum, rosenbachianum, jesdianum, hirtifolium* and often a shorter allium than *A. aflatunense* which has not been identified. To add to the confusion *AA. giganteum, elatum* and *macleanii,* other tall drumsticks, are involved in synonymy problems.

The publication in 1986 of Volume 1 of the *European Garden Flora* contains an entry listing alliums, with the authority of W.T. Stearn and E. Campbell. Gardeners who hope to establish the identity of their 'drumsticks' will find the botanical descriptions here or in *Flora USSR*.

Asia Minor, Central and South-West Asia are the origins of the alliums listed above. Many of the bulbs commercially available are surprisingly resilient. Ranging between 30–150 cm (12–60 ins), but more frequently attaining 45–60 cm (18–24 ins), they can be grown in sunny borders with good drainage. When happy they will seed, allowing a stock to be built. The spherical heads of mid to deep lilac blend happily into most colour schemes, provide a pleasing foil for informal plantings and dry for indoor use. Seed is usually formed in abundance, taking 3–5 years to attain flowering size. Most drumsticks flower spring and early summer, requiring varying degrees of rain protection.

A. aflatunense has ovoid bulbs with papery coverings. The 6–8 leaves are strap-shaped, smooth-margined, basal and much shorter than the flowering stem. Slight ribbing can be seen on the 80–150 cm (32–60 ins) stem. The spherical, many-flowered head is light violet, 7–10 cm (2½–3½ ins) wide. Darker-veined tepals have stamens projecting slightly beyond their tips. HZ 4

SOURCES *Fl USSR*, 209; *Eur Garden Fl*, 74; *Hortus III*. Illus: Hay & Synge, *Dictionary Garden Plants*.

A. akaka
Gmelin 1830

Coming from Iran, Turkey and the Caucasus (USSR), where it grows in dry stony slopes, at 1,500–3,000 m (4,500–9,000 ft), this small plant requires a bulb frame or pot culture, with summer rest. *A. akaka* is becoming available from specialist bulb nurseries, making it easier to obtain than some of the other attractive alliums from Iran, for example *AA. mirum, bucharicum* or *protensum*.

The hemispherical umbel, 3–10 cm (1–4 ins) in diameter, carries numerous pale pink or whitish flowers with a central deeper vein, held on a stocky stem 8–30 cm (3½–12 ins) tall. The leaves are more opulent: glaucous and wide, up to 20 × 6 cm (8 × 2½ ins), they resemble those of the better-known *A. karataviense* or perhaps a Kaufmanniana Tulip. April/May. (Pl. p. 40.) HZ 5

SYNONYM *A. latifolium*.

SOURCES *Fl USSR*, 190; *Fl Iranica*, 98; *Fl Turkey*, 120. Illus: Mathew, *Smaller Bulbs*; Rix & Phillips, *The Bulb Book*.

A. albidum (= whitish)
Fischer ex Bieberstein 1819

Several slender, greyish, cylindrical bulbs are found clustered on a horizontal rhizome. Arising from them are ribbed stems, 10–30 cm (4–12 ins), with 5–9 shorter, basal, thready leaves. Below the hemispherical or shuttlecock-shaped umbel, 1.5–2.5 cm (⅗–1 in) in diameter, the spathe splits into 2 valves. Many white or yellowish flowers, with very slightly exserted or equal stamens, grow on 8–15 mm (½–¾ in) stalks. These are unusual in being ribbed and rough with minute teeth.

In *A. albidum* subsp. *albidum*, which includes *A. flavescens* and *A. ammophilum*, the pedicels are two or three times longer than the yellowish-white flowers.

In *A. albidum* subsp. *caucasicum* the whitish flowers, sometimes found with a pink tinge, are held on pedicels 1½–2 times as long as the tepals.

The subspecies' differences account for the slight variance in the 'vital statistics' of *A. albidum* in different floras. Ranging from the Balkans through the Caucasus, Iran and Turkey, they are found in steppes, sandy and rocky places, flowering July. *A. albidum* is easily grown from seed and enjoys sunny, well-drained areas in the garden. It is a pleasant if not spectacular little plant. HZ 5

SYNONYMS *A. ammophilum, A. flav-escens, A. angulosum* var. *caucasicum.*

SOURCES *Fl Eur,* 6; *Fl USSR,* 38; *Fl Turkey,* 3; *Eur Garden Fl,* 5; Grey-Wilson & Matthew, *Bulbs.* Illus: Reichenbach, *Icones Florae Germanicae,* 10: 499.

A. albopilosum (= whitehaired)
C.H. Wright 1903

Described in the *Gardeners Chronicle* in 1903, but an earlier account dated 1884 necessitated name changing. Therefore, details of this very popular and spectacular allium will be found under *A. christophii.*

A. altaicum (= from the Altai)
Pallas 1773

The stout, hollow stem rising to 30–70 cm (12–28 ins), narrows just below the umbel and is sheathed with leaves to almost half its length. These too are hollow, smooth and almost as long as the stem, one or two pairs being present. The spherical head is packed with yellowish-white florets, the whole being rather small, 1.5–5 cm (3/5–2 ins) wide, in proportion to the stem. While not very decorative in the garden, bees love the plant; a good reason for growing the species. The bulbs have a reddish-brown covering and are attached to an oblique rhizome.

Coming from rocky areas of Siberia and Mongolia, plants are perfectly hardy and can be grown in the open garden. Most probable ancestor of *A. fistulosum,* the Welsh Onion, *A. altaicum* flowers in July/August. HZ 1

SOURCES *Fl Eur,* 21; *Fl USSR,* 87; *Hortus III.*

A. altissimum (= tallest)
Regel 1884

Despite the similarity in names this is a very different plant from *A. altaicum,* being a purple 'drumstick'. The large heads are about 8 cm (3½ ins) in diameter, packed with broad, starry flowers on long unequal flowerstalks and distinguished by the purple filaments. The bulbs are globular, have greyish, papery coats, while the stem is 80–150 cm (32–60 ins), with basal sheaths of 4–6 leaves, shorter than the stem.

Flowering in April/May, plants require sharp drainage and a warm site. Iran, Afghanistan, Central Asia. HZ 4

SOURCES *Fl USSR,* 210; *Hortus III;* Mathew, *Large Bulbs.* Illus: *Flora Iranica.*

A. amabile (= lovely)
Stapf 1931

Possibly a synonym of *A. mairei.* One of the daintiest of the genus, *A. amabile* was introduced by George Forrest from northwest Yunnan, south-west China, in 1922, from stony alpine meadows at an altitude of 4,200 m (12,600 ft). Despite this, it revels in wet, acid soil in north-west England at 230 m (700 ft). The fine grassy foliage, 10 cm (4 ins) tall makes a splendid slugs' hors-d'ôeuvre, but if the seedlings are grown on in a pot until the clump is approx. 4 cm (2 ins) diameter before planting out, survival is reasonably assured. In July the narrow, maroon-pink flowers appear and may continue until the middle of August.

Flowering to 12–16 cm (5–6 ins), with the leaves still green at maturity, *A. amabile* is a delightful small plant for moist areas of the rock garden. While self-seeding has been no problem in our very damp garden, seed deliberately sown in autumn has flowered the following summer. The slender bulbs on a short rhizome are easily divisible. The umbel carries only a few flowers, 3–6, but the clump as a whole makes an effective patch of colour.

While botanically *A. amabile* is considered hardly distinct from *A. mairei,* which

A. amabile in Hartsop, Cumbria (Author)

takes precedence, gardeners will have no difficulty distinguishing the two. Though both are delightful alliums for the rock garden, plants known as *A. mairei* in commerce are a paler pink than *A. amabile*.

SOURCES *Fl R Pop Sin*, 42; *Hortus III*; Synge, *Collins' Guide to Bulbs*. Illus: *Bot Mag*, 9257, 1931; Rix & Phillips, *Bulb Book*.

A. amethystinum
Tausch 1828

The almost spherical umbel, 2.5–6.5 cm (1–2½ ins), carries many purple flowers, the outer on drooping flower stalks, the inner standing upright. Each globular bulb carries a stem 30–120 cm (12–40 ins) tall, which may have a reddish tinge in the upper half. Hollow leaves sheathe the lower stem and, like many other alliums, have withered by flowering time. The hairy stamens are longer than the tepals.

Flowering in May to July on cultivated ground and rocky places in Eastern and Central Mediterranean regions, *A. amethystinum* can be grown outside in a sunny, well-drained bed. HZ 4–5

SYNONYMS *A. segetum*, *A. rollii*, *A. descendens* auct non L.

SOURCES *Fl Eur*, 96; *Fl Turkey*, 105; Grey-Wilson & Mathew, *Bulbs*. Illus: Polunin, *Fl of Greece & Balkans*.

A. ampeloprasum
(= the onion of the vineyard)
Wild Leek, Great Headed Garlic
Linnaeus 1753

Many yellowish bulblets are produced by the rounded bulb. The stem, 40–180 cm (14–60 ins), is sheathed on the lower part by linear leaves which have withered by flowering time. Each 5–10 cm (2–4 ins) wide, large umbel is packed with white, pink or reddish flowers or sometimes with an admixture of bulbils, with a single spathe. The stamens slightly protrude beyond the tepals which are distinguished by hairy nodules on the outer surface.

Flowering period is throughout summer. Southern Europe, North Africa, Asia Minor into the Caucasus, on dry, rocky or sandy grounds and bordering cultivated land. HZ 2

SYNONYM *A. halleri*.

A. porrum (qv)

The cultivated Leek is a cultigen of *A. ampeloprasum*.

A. ampeloprasum var. *babingtonii*

Has many bulbils in the head and is a rare native found along the Atlantic seaboard, often in drainage ditches, in western Ireland and Cornwall. The plant may be a relic of early cultivation on the sites of old monasteries and settlements.

As a dried head for winter decoration, it has a certain angular charm and a not displeasing hint of garlic lingers around the display. At the back of an herbaceous border the towering heads are quite statuesque without requiring staking. As the bulbils are not too easily dislodged until well ripened, the plants can be grown without undue risk of spreading far and wide.

As a further, amusing feature, the bulbils

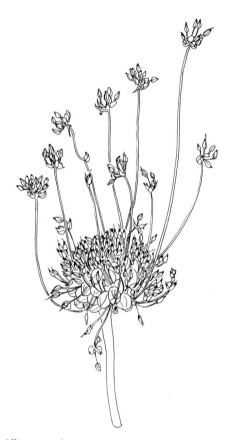

Allium ampeloprasum var. *babingtonii*

themselves sprout while still growing in the flower head, sending up 'grandchildren' flowers. Possibly, under ideal conditions, this fecundity could continue for several generations, the flower head eventually collapsing under a mass of progeny. E.A. Bowles allowed bulbs in his borders, setting an unimpeachable precedence for their admittance into the garden (*My Garden in Summer*).

Commemorates C.C. Babington, 1808–1895, the British botanist.

SYNONYM   *A. babingtonii*.

A. ampeloprasum var. *bulbiferum*.
Found in the Channel Islands and western France.

AA. bourgeaui, pardoi, polyanthum, pyrenaicum, commutatum and *scaberrimum*
These are several of the species related to *A. ampeloprasum*, but showing botanical differences of little horticultural interest.

SOURCES   *Fl Eur*, 76; *Fl USSR*, 175; *Fl Turkey*, 76. Illus: *Bot Mag*, 1385 (1811); Polunin, *Concise Flowers of Europe*.

A. amplectens
(= embracing, clasping the base)
Narrow-leaved Onion, Paper Onion
Torrey 1857
Globular heads bear numerous white or pink flowers, which become paper-thin after fertilisation, each carried on a longish flower stalk. The yellow or purplish stamens are shorter than the tepals. Stems are 10–40 cm (4–16 ins), sheathed by 2–4 early-withering leaves. Ovoid bulbs with brownish coats present with transverse, broadly V-shaped meshes in vertical rows, beneath which the inner coats are reddish or off-white.

Flowering in March–June in open fields and on hillsides from Vancouver and San Juan Islands, Washington to west of the Sierra Nevada, California, plants need a summer rest and will grow best in a bulb frame or pot. HZ 2

SYNONYMS   *A. attenuifolium, A. occidentale, A. monospermum*.

SOURCES   *Hortus III*; Munz, *California Flora*. Illus: Hitchcock *et al.*, *Vasc Pl Pac N W*; Cronquist *et al.*, *Intermountain Flora*, 13.

A. angulosum
Linnaeus 1753
A large group of alliums found both in the wild and in cultivation resemble *A. senescens*. They are difficult to differentiate

and the minutiae of botanical assessment will interest the average grower less than their value in the garden. *A. angulosum* and *A. suaveolens* (suaveolens = fragrant) have keeled leaves to separate them from others in the group, which have the under-surfaces flat or rounded.

A plant currently available in Europe and North America as '*A. pyrenaicum*' is probably *A. angulosum*. The true *A. pyrenaicum* is a relative of *A. ampeloprasum*, having white flowers on a Leek-like plant.

A. angulosum has slender, elongated bulbs clumped around a rhizome. The basal leaves, 4–6 in number, are sharply keeled on the undersurface, rich green and shiny, with blunt tips. Unlike many an-other *Allium*, they remain glossy and un-withered throughout the flowering period, adding to its value as a garden plant. The stout and fleshy stem is 20–45 cm (7–15 ins), flattened at the top to appear distinctly 2–cornered and crowned by a hemispherical umbel of pale lilac flowers, 2.5–4.5 cm (1–2 ins) wide. The anthers approximate the tepal length and darken with maturity.

Coming from damp grasslands, from Europe through to Siberia, this bonehardy *Allium* is totally reliable even in acid, cold soils, indeed it requires moist conditions to flourish. The absence of aggressive onion odour makes it a good cut flower. (Fig. p. 24) June–August. HZ 1

SOURCES *Fl Eur*, 1; *Fl USSR*, 37; *Hortus III*; Grey, *Hardy Bulbs*; *RHS Dictionary*. Illus: Redouté, *Liliacées*, 5: 281 (1809); Grey-Wilson & Mathew, *Bulbs*.

A. anisopodium
(aniso = unequal; podium = base, foot)
Ledebour 1853

A rhizomatous cluster of bulbs support 20–40 cm (8–16 in) stems, with 2–3 slightly shorter, slender, semi-cylindrical leaves. Numerous rosy, bell-shaped flowers on longish, unequal stems form an umbel slightly shuttlecock in shape. (Grey describes the tepals as white.)

Growing on dry slopes and sand from Mongolia to Japan and China, flowering in May, plants survived on a raised bed in the wet, cool climate of Cumbria for several years. HZ 2

SOURCES *Fl USSR*, 51; Grey, *Hardy Bulbs*; *Hortus III*; *Fl Rei Pop Sin*, 45.

A. ascalonicum
Linnaeus 1759

This is a name that over the years has been incorrectly applied to the garden Shallot. The original plant described by Linnaeus as *A. ascalonicum* is probably the same as *A. hierochuntium* Boissier. The Shallot is considered to be a variant of *A. cepa*.

Named after Askalon (modern Askulan), ancient city near Jerusalem, birthplace of Herod the Great.

A. atropurpureum
(atro = black or dark-coloured)
Waldstein & Kitaibel 1800

Another pleasant 'drumstick' for the garden. The globular bulbs have membraneous coverings. There are 3–7 basal leaves, less than half the length of the stem, which rises to 40–100 cm (13–33 ins). Flower heads, 3–7 cm (1¼–2¾ ins) in diameter, with bi-valved bracts, carry numerous maroon-purple, starry flowers set with purple-anthered stamens shorter than the tepals.

Found in cultivated ground and dry open spaces throughout the Balkans, June/July. Grows well in a sunny, well-drained bed. HZ 4

SYNONYM *A. nigrum* var. *atropurpureum*.

SOURCES   *Fl Eur*, 105; *Fl Turkey*, 123;
Eur Garden Fl, 67; *Hortus III*. Illus:
Mathew, *Larger Bulbs*.

A. atrorubens
S. Watson   1871

This small onion from America, despite its
name, is a complete contrast to the similarly named 'drumsticks', so finds its way
into this list of species to avoid confusion
with them.

The ovoid bulbs are usually not on a
rhizome and have red-brown coats with
indistinct netted surfaces. Cylindrical
stems, 6–15 cm (2–5 ins), carry many-
flowered umbels studded with reddish-
purple, rather stiff, pointed tepals. Its
pedicels are slightly longer than the
flowers, while the stamens are two-thirds
the tepal length capped with yellow or
purple anthers. Prominent crests crown the
ovary, the stigma is unlobed and nectaries
are absent. Solitary cylindrical leaves over-
top the stem.

A. a. var. inyonis

Has pale tepals with dark mid-veins (Inyo
Co., California).

This small onion can be found blooming
through May and June on dry hillsides,
1,700–2,300 m (5,000–7,000 ft) in eastern California and Nevada. Recording that
A. atrorubens grows in the Avawatz
Mountains, Death Valley, gives more than
a broad hint that cultivation in a bulb
frame or pot with summer dormancy will
be more successful than in a damp garden
bed. HZ 4

SOURCES   *Hortus III*; Munz, *California
Flora*; Grey, *Hardy Bulbs*. Illus: Cronquist
et al., *Intermountain Flora*, 9; Rickett,
Wildflowers U S, Vol. 4.

A. atroviolaceum
Boissier   1846

A black-purple 'drumstick' found enliven-
ing the gold of Iranian wheatfields. The
ovoid bulbs, with membraneous coverings
becoming fibrous, produce numerous,
small, yellow-brown bulblets. Tall stems,
60–100 cm (20–33 ins), are clothed over
the lower halves with up to 6 linear leaves
minutely toothed on the margins, all of
which have withered by flowering time.
A long-beaked, single spathe enfolds the
densely packed umbel. This is spherical,
3–6 cm (1–2 ins) in diameter, the flowers
dark purple (rarely greenish), bell-shaped,
with long stamens and shorter style.

Rather similar to *A. scorodoprasum*
subsp. *rotundum*, the latter may be differentiated partly by its short stamens,
included within the flower.

Flowering in June–August on dry slopes
and cultivated fields, *A. atroviolaceum* is
found from south-east and east Central
Europe, as far afield as Central Asia and
Afghanistan. Easily grown in the garden in
good drainage and sun. HZ 3

SYNONYM   *A. ampeloprasum* var. *atroviolaceum*.

SOURCES   *Fl USSR*, 173; *Fl Eur*, 79; *Fl
Iranica*; *Fl Turkey*, 83. Illus: Wendelbo,
Tulips and Irises of Iran.

A. babingtonii

See *A. ampeloprasum* var. *babingtonii*.

A. barszczewskii
Lipsky   1900

One or 2 rather conical bulbs with brown,
netted coverings appear attached to an
oblique rhizome. Many small bulblets are
found approximately 1 cm (2/5 in.) higher
on the part of the stem below ground. This
ranges between 20–60 cm (8–20 ins),
clothed on its lower third by leaf sheaths.
Shorter than the stem the leaves are finely
linear, numbering 3–5 and withered by
flowering. (Pl. p. 9.)

The stem may often grow in hooped
form, straightening by flowering time. An

almost spherical umbel 2 cm × 2.5 cm (³⁄4 × 1 ⅛ ins), is packed with pinky-lilac, bell-like flowers bearing a deeper stripe. Plants with white flowers have also been described. The stamens are shorter than the tepals, the outer being lilac and the inner cream, so that the centre of each floret appears lighter than the outer edge giving an attractive, two-toned appearance to the flower head.

Originating on stony slopes in Central Asia, spanning lower and upper mountain zones, in flower from May to August, this onion has survived two mild winters in one of our raised gravelly beds, but might be safer in a bulb frame. HZ 3

According to *Flora USSR*, *A. barsz-czewskii* may be confused with some plants identified as *A. tataricum*.

SOURCES *Fl USSR*, 27; *Fl Iranica*, 8.

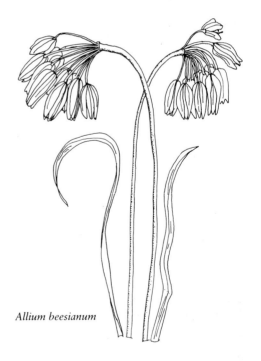

Allium beesianum

A. beesianum
W.W. Smith 1914

A. beesianum is one of the aristocrats of the Alliaceae and an exceptionally attractive bulb for the rock garden, flowering as it does in late August–September. Blue tones are always welcome in the garden, so are flowers of classic grace.

Collected by George Forrest in 1910 in the Lichiang Range of north-west Yunnan, southern China, it was named after the firm of Bees in north-west England. Mr Bulley, head of the firm and commissioner of several plant hunting explorations, introduced many fine plants through his nursery. His reward was to have several plants named for him and his firm. His garden has remained in horticulture, being the present home of Liverpool University's Botanic Garden at Ness.

Narrowly cylindrical bulbs with netted fibres arise from thready rhizomes. Two–4 leaves sheathe the lower third–half of the stem. Purplish at the base, tapered and erect, they are shorter than the 20–35 cm (8–14 ins) stem. Each umbel holds

6–12 flowers, bright blue, on drooping 7–10 mm (¼–½ ins) flower stems. The flowers are tubular bells, 11–17 mm (½–¾ ins) long and 2.5–5 mm (⅛–⅕ ins) wide, with blue stamens shorter than or just level with the segment tips. Both pale and bright blue forms are commonly seen and while other colours have been reported—purple-blue, white, magenta and possibly a purple-violet—none of the latter is in general cultivation.

Indeed it is quite difficult to obtain *A. beesianum* at all. Many of the plants erroneously in commerce are *A. cyaneum*, cobalt blue, flowering in late summer, in which the stamens are much longer than the tepals.

Other blues are *A. kansuense*, probably a form of *A. sikkimense*, and *A. sikkimense* itself (qv), smaller plants flowering earlier in the year, May–June. Although having stamens reaching to the tepal ends, there should be little reason to confuse *A. beesianum* with these. From seed exchanges, however, the most common error is seed gathered from *A. cyathophorum* var. *farreri* (qv), a small robust, purplish rock plant.

Growing on stony areas at 3,000–4,000 m (9,000–12,000 ft), cultivation is easy in well-drained, sunny beds that are not too dry. Flourishing groups have been noted at Threave in south-west Scotland. HZ 4

SOURCES *Fl R Pop Sin*, 33; *Hortus III*. Illus: *Bot Mag*, 9331 (1933); Synge, *Collins' Guide to Bulbs*; Synge & Hay, *Dictionary Garden Plants*; Bacon, *AGS Bulletin*, Vol. 49, no. 2, p. 158.

A. bidwelliae
S. Watson 1880

Synonym of *A. campanulatum*. No longer given species status, the name appears quite frequently in the literature.

SOURCES *Hortus III*; *RHS Dictionary*; Wilder, *Hardy Bulbs*. Illus: Rickett, *Wildflowers U S*, Vol. 6 (*A. campanulatum*).

A. bisceptrum (= two-sceptred)
Aspen Onion
S. Watson 1871

A small onion from the Great Basin and along the eastern side of the Cascades down to the Sierra Nevada. As the common name suggests, this is at home in aspen groves. Greyish, round bulbs, surrounded with numerous bulblets, support several round stems 10–13 cm (4–5½ ins) high. The 2–3 flat leaves are 3–10 mm (⅛–⅖ ins) wide and as long as the stem. Two-valved, pointed bracts accompany the multi-flowered, open umbel whose pale to dark rose tepals are sharply pointed with greenish bases. Pairs of saw-toothed, triangular appendages sit on top of the ovary lobes.

Common in meadows, moist banks as well as aspen groves at 2,000–3,200 m

(6,500–9,500 ft) and montane coniferous forests, from south Oregon to Nevada, Utah and Idaho in May–July. Sunny, well-drained beds. HZ 4.

SYNONYMS *A. tenellum*, *A. bullardii*, *A. palmeri*.

SOURCES Munz, *California Flora*, 4; *Hortus III*. Illus: Niehaus & Ripper, *Field Guide to Pacific States Wildflowers*; Hitchcock *et al.*, *Vasc Pl Pac N W*; Cronquist *et al.*, *Intermountain Flora*, 10.

A. bodeanum
Regel 1875

Described in *Flora USSR* as being very close to *A. christophii*, Per Wendelbo in his book *Tulips and Irises of Iran* calls it one of the more remarkable species of its genus. As it comes into cultivation it should be a very popular addition to the garden flora.

The broad, basal leaves are blue-green with a large, spherical head of flowers sitting on a short stem, 10–20 cm (4–8 ins) in height. Each individual floret is lilac and star-shaped.

The plant comes from gravelly slopes in the steppes of Central Asia and Iran, flowering in May. Rather similar species are *AA. akaka*, *haemanthoides* and *elburzense*. Many of these broad-leaved alliums are relished by the local inhabitants. Few western vegetable gardens will be able to provide the hot, dry situations to make them an easy crop outdoors! HZ 4

SOURCES *Fl USSR*, 193. Illus: Wendelbo, *Tulips and Irises of Iran*.

A. bolanderi Bolander's Onion
S. Watson 1879

A tiny Allium found in the Siskiyou Mountains from southern Oregon into California, usually growing in a clay soil baking hard in summer. Flowering May–

July, *A. bolanderi* favours clearings in chaparral, Yellow Pine forests and woodland below, 1,000 m (3,000 ft).

The bulbs are shallow-growing, greyish-white, very small, 5–10 mm (⅕–⅖ ins). The slender stem, 10–20 cm (4–8 in) supports a head of small, bell-like flowers, rose-white to pink, the long, narrow tepals 8–15 mm (⅓–¾ ins). The stamens are half as long as the tepals with pale yellow anthers, the stigma slightly lobed. Two–4 thready leaves, shorter than the stem, have mostly withered by flowering time.

A. b. var. stenanthum
(syn. A. stenanthum)

Is somewhat taller than the type with narrower tepals, usually white. Found in open Pine woods, Humboldt Co., California.

A. bolanderi would be lost in the open garden. Plants seen in the Siskiyou Mountains were only 5 cm (2 ins), slight enough for a small trough; however, other specimens grown in pots were rather taller while retaining airy lightness. Raised bed or bulb frame. HZ 3

SYNONYM *A. stenanthum.*

SOURCES Munz, *California Flora;* Grey, *Hardy Bulbs.* Illus: Abrams, *Illus Fl Pac States,* 1, no. 42.

A. brevistylum Shortstyle Onion
Watson 1871

Slender bulbs terminate in a thick rhizome with a membraneous covering. Still fresh and green at flowering, the leaves are shorter than the stems, which are 20–60 cm (8–24 ins), slightly flattened and topped with 2 persistent spathe valves. Each flower head may carry 7–15 individual flowers on rather slender stalks. These become thickened and curved as the seeds mature. The pink tepals have pointed tips and a slightly thickened mid-rib,

Allium brevistylum

10–13 mm (⅓–⅔ ins) long. Yellow anthers top the included stamens, the style is cleft in three.

A. brevistylum flourishes in swampy meadows and along stream sides in the Rocky Mountains at medium and high elevations. None the less, it was to be seen at the top of a rock slope at 3,000 (10,000 ft) on Snowy Pass, Medicine Bow Mountains, Wyoming in July. A large snow patch was melting at the base some 6 m (20 ft) below, the drainage was very sharp and the remainder of the short summer would be hot, dry and windy. Flowering in June–August, this is a common *Allium* in the Bear Tooth Mountains and Yellowstone National Park, Wyoming—indeed from Montana, and Idaho through Utah and Colorado.

A. brevistylum may be confused with *A. validum,* the latter differs in its stamens which protrude well beyond the tepal edges.

Garden culture should not be difficult if the plant's need for moisture is remembered, with an open site but drier conditions in the winter. HZ 2

SOURCES *Hortus III*; Rydberg, *Flora of Rocky Mts.* Illus: Hitchcock *et al.*, *Vasc Pl Pac N W*; Rickett, *Wildflowers U S*, Vol. 6; Shaw, *Plants of Yellowstone & Grand Teton National Parks* (excellent photo).

A. breweri Brewer's Onion
S. Watson 1879

A plant of the Coast Ranges of California, this is now considered a synonym of *A. falcifolium* (qv).

SOURCES Munz, *California Flora*; Grey, *Hardy Bulbs.* Illus: Abrams, *Ills Fl Pac States*, 1, 12.

A. bucharicum
Regel 1884

Subglobose bulbs, 1–3 cm (²/₅–³/₄ ins) in diameter, with dark, papery coats, produce a dumpy stem 10–30 cm (4–12 ins). Three–6 basal leaves, 7–20 mm (½–³/₄ ins) wide, with minutely toothed margins, are a little longer than the scape. The hemispherical, occasionally spherical, umbel has numerous stellate flowers on unequal stalks, 2–6 cm (³/₄–2⅓ ins) long. A greenish-purple vein runs down the centre of the white tepals, which become erect and rigid after flowering.

A. bucharicum belongs to an interesting group of bulbs from Iran, Turkey and Afghanistan. Some were described at the end of the last century but only found their way into cultivation in Britain around 1962, many through the collections of Rear Admiral Paul Furse.

Vvedensky, in *Flora USSR*, 1935, included *A. bucharicum* as a synonym of *A. schubertii*. By 1963 he separated the two,

describing *A. bucharicum* as having dull rose tepals striped with dingy purplish veins, from the Pamir-Alai and *A. schubertii* as originating in the Aral-Caspian area. Hedge & Wendelbo described plants with white tepals and greenish-purple nerves that corresponded with Regel's description, (this is how *A. bucharicum* appears in Wendelbo's account of Alliaceae in *Flora Iranica*, 1971.)

The very large flower heads on short scapes are quite spectacular, even if the colour of the tepals may be somewhat dull. Plants are difficult to obtain; propagation is primarily by seed. In turn this is not easy to come by and, when grown, too often the plants are a great disappointment, by dint of being another species altogether. The latest to flower in my collection of seedlings turned out to be *A. paniculatum*! *A. bucharicum* appeared in the RHS Lily Group exhibit of *Allium* in 1986 under the number PF 6268.

Growing at 1,500–2,100 m (4,500–6,300 ft) in Afghanistan, cultivation when available would be most certainly bulb frame or alpine house with summer dormancy. HZ 4

SOURCES Grey, *Hardy Bulbs*; *Fl USSR*, 219 (as *A. schubertii*). Illus: *Fl Iranica*, 130.

A. bulgaricum

Strictly speaking this plant should no longer be included in an enumeration of *Allium*, having been allocated to the genus of *Nectaroscordum*. However, so popular are the seed heads with flower arrangers, while the old name is still familiar to many gardeners *N. siculum* subsp. *bulgaricum* deserves recording here.

The tall stems, 50–125 cm (20–48 ins), carry flowers of greenish-white, the exteriors tinged pink, with green mid-veins and patches of red towards the inside of the flower base. Before fertilisation the flowers hang their heads, then the seed heads stand proudly erect and furthermore dry well to pale parchment. Easily grown even in damp clay soil in north-west England.

See also *A. siculum.*

SOURCE Illus: Rix & Phillips, *Bulb Book.*

A. caeruleum
(= the blue of the heavens)
Pallas 1773

An attractive flowering plant easily obtained through popular bulb catalogues. The 3–4 cm (1–1½ ins) umbel is almost globular on a 20–80 cm (9–24 ins) stem. The narrow, 3-cornered leaves have withered by flowering time and, although the flowering heads are quite small, a thick planting can produce an effective show. Medium blue tepals have a darker mid-vein, the stamens are about the same length. Plants are easily grown in a dry spot, in May and June. Grows in the Russian steppes and in inland salt marshes. HZ 3

The *Fl USSR* uses the name *A. coeruleum* and comments that it can hybridise with *A. caesium* where they share habitats. Plants occurring with bulbils in the head have been called *A. coeruleum* var. *bulbilliferum* Ledebour, *A. coeruleum* var. *viviparum* and *A. viviparum.*

SYNONYM *A. azureum.*

SOURCES *Fl Eur*, 54; *Fl USSR*, 129; *Hortus III.* Illus: Rix & Phillips, *Bulb Book*; Synge, *Collins' Guide to Bulbs.*

A. caesium
(= lavender-blue, or pale blue tinged with grey)
Schrenk 1844

Very similar to *A. caeruleum* but not as easily available from bulb firms. The

A. caeruleum (Author)

flower head is a little greyer and E.B. Anderson in *Dwarf Bulbs for the Rock Garden* wrote that it flowered two weeks later than *A. caeruleum*. The leaves differ from *A. caeruleum*, being hollow and semicylindrical. Both of these slender blue alliums may have bulbils in the head. A white form has been recorded.

Mark McDonough writes of a form with umbels 9 cm (3½ ins) in diameter with a fragrant perfume. He also recommends picking the heads of both these blue species before they become dry, when the blue tones persist for flower arrangement, albeit paler.

Found from Siberia to Central Asia in the steppe belt, semi-deserts and mountain sides to 2,000 m (6,000 ft). A dry, hot corner, bulb frame or alpine house is required for cultivation. HZ 3

A. callimischon var. *haemostictum* (Author)

SOURCES *Fl USSR*, 130; *Hortus III*. Illus: Polunin, *Flowers of the Himalaya*, pl. 121; McDonough in *ARGS*, vol. 42, no. 3, 1984.

A. callimischon
(calli = beautiful; mischon = husk/shell)
Link 1835

This very small onion flowers in September through to November. While it has proved hardy in a dry, raised beds in southern England, it can be more easily admired in an alpine house or frame. HZ 5

The leaves appear in autumn, a short flowering stem, 9–38 cm (3½–15 ins), grows in spring and then appears to wither with the leaves, only to produce flowers the following autumn. Tidy growers should avoid the temptation to neaten their plants.

A. c. subsp. *callimischon*

From the Peloponnese. The clump of 8–25 flowers, white with a distinct reddish-brown vein, are unspotted.

A. c. subsp. *haemostictum*

From Crete. Has dark red spots (Gr. haima = blood).

While this would be an unspectacular plant in midsummer, few flowers are in bloom so late in the year that *A. callimischon* is doubly welcome (Fig. p. 56).

SOURCES *Fl Eur*, 50; *Fl Turkey*, 24. Illus: Rix & Phillips, *Bulb Book*; Polunin, *Fl of Greece & Balkans*.

A. calocephalum
(calo = beautiful, cephalum = head)
Wendelbo 1966

This distinctive *Allium* has both fertile and infertile flowers, the latter on the outside of the flower head have very long tepals, 3 cm (1¼ ins). Seed has not yet started to appear in seed lists and plants are not apparently available in commerce. Indeed, it is possible that the species is not in cultivation in Great Britain even though accounts of the plant were available around 1970, and herbarium material at Kew from 1934 shows that it was grown many years before by Ludlow-Hewitt.

Originating in the Middle East, flowering in spring, it would be an attractive plant for the rock garden. Dr Jack Elliott, one of the most experienced and successful bulb growers, received plants from Admiral Furse, which quietly faded away leaving only colour slides as a tantalising memory. Cultivation will obviously be no easy matter unless more amenable forms are introduced in years to come.

The tepals are creamy-lemon, long and wispy with conspicuous ovaries of chocolate hue. The appearance is more reminiscent of a large, pale *Globularia* or *Jasione*, than of a member of the Onion family. Excellent photographs appear in *Flora Iranica* and in the *Journal of the RHS*, April 1970, Part 4, contained in an account of ornamental Onions by the late Keith Blanchard.

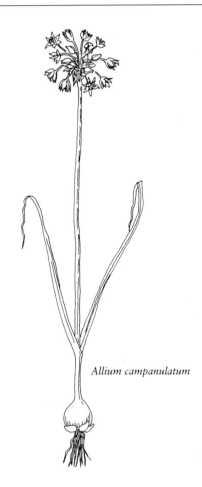

Allium campanulatum

Allium calocephalum

A. *campanulatum* Sierra Onion
S. Watson 1879

The ovoid bulbs, 15–20 mm (¾–1 ins) long, have brown outer coats over red-tinged inner linings, from which rise clusters of short-stalked bulblets either directly or on short, thin rhizomes. The 2 leaves are slightly shorter than the solid 15–30 cm (6–12 ins) stems and are often withered by the time the flowers are opening. Occasionally there may be 2 or more stems from a single bulb. The lax, hemispherical umbel has 2 bracts, numerous rose-purple or sometimes white flowers, with cup or star-shaped tepals on long stalks 15–20 mm (¾–1 ins) which become rigid as the seed ripens.

A. campanulatum is found growing in open coniferous forests in California, Nevada and eastern Oregon, in dry soils at medium and high elevations, flowering in May–July. In the garden it will require a summer rest, bulb frame or alpine house. HZ 4, 5

A certain amount of confusion occurs in the literature. *A. bidwelliae* has been described as being a tall version of *A. campanulatum*; flower colour is also often stated as pale pink.

SYNONYMS   *A. bidwelliae, A. austinae.*

SOURCES   Illus: Hitchcock *et al., Vasc Pl Pac N W*; Rickett, *Wildflowers U S*, Vol. 4.

A. canadense Wild Garlic, Meadow Leek, Rose Leek, Canada Garlic
Linnaeus   1753

A. canadense is a variable species, with many forms producing bulbils in the head. Few gardeners would wish to grow it in the flower beds. Today it is replaced in the vegetable garden by other alliums. Back in 1674, Marquette and an exploratory party journeyed from Green Bay to the modern-day site of Chicago using *A. canadense* as their main food supply. It had a traditional use as an Indian food source and at a later date was popular as a stew flavouring with Maine lumbermen. As a pickling onion, the flavour is said to be superior.

The stem is about 30 cm (12 ins) with 3 or more leaves; shorter than the scape. Unlike Garlic, the spathe valves number 2–3. Flowers when present are pink or white. The bulbils are stemless. Several varieties are recorded.

A. c. var. *canadense*
Most or all the flowers are replaced by bulbils.

A. c. var. *fraseri*
The flower stalks are stout and the florets white.

A. c. var. *mobilense*
The flower stalks are slender and the flowers pink.

A. c. var. *lavandulare*
Is strictly floriferous.

A. mutabile
Was once used to denote the species carrying flowers only, but has also been misapplied to *A. drummondii*.

Including all the varieties, the range is wide across North America, in open woods and prairies, flowering May–July. HZ 1

SYNONYMS   *A. continuum, A. canadense* var. *ovoideum, A. c.* var. *robustum, A. longicaule.*

SOURCES   Ownbey, *Genus Allium in Texas; Hortus III.* Illus: Rickett, *Wildflowers US*, Vol. 6; Niehaus & Ripper, *Field Guide to S–W US & Texas.*

A. caput medusae
Airy Shaw   1931

With a lovely illustration in *Hardy Bulbs*, Grey introduces a bulb with mythical connections. Medusa, one of the Gorgons of the Greek pantheon, had the power to turn any unwary onlooker to stone. Slain by Perseus, her head mounted on his shield, her petrifying influence was not abated by death. Atlas became a mountain, serpents were spawned, sea monsters granitised and a mere 200 warriors turned to stone. A further banqueting hall of diners was immobilised before Perseus very wisely turned this dangerous head over to Athene, goddess of wisdom.

Our *Allium* was found by Frank Kingdon Ward in 1919 in a bamboo forest in the valley of Chawng-maw-hka in the heights of Upper Burma. The flowers were nodding, reddish-purple, flowering at the end of July. The bulbs collected by Captain

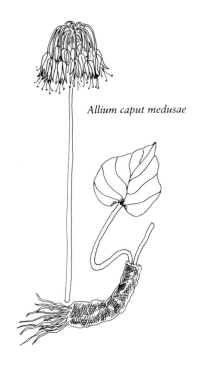

Allium caput medusae

Kingdon Ward failed to survive and seed was not available, so for the present this colourful onion has to remain in the realms of myth.

The drawing in *Hardy Bulbs* was produced from dried material preserved in Edinburgh.

A. cardiostemon
(= with heart-shaped stamens)
Fischer & Meyer   1840

From Iran and the Caucasus comes this 20–40 cm (9–18 ins) tall *Allium*, with a hemispherical umbel, 3–5 cm (1¼–2 ins) in diameter, packed with blackcurrant-purple flowers. The smooth stem is clothed at the base with 4–6 shorter leaves which are still fresh at flowering time.

C.H. Grey suggested that this attractive small plant was easily grown in any well-drained soil, flowering in June. Cultivation in an alpine house or frame would be safer as a dry summer rest is required. The rich,

deep colour makes *A. cardiostemon* quite a striking pot plant. HZ 4

SYNONYMS   *A. atriphoeniceum, A. trilophostemon.*

SOURCES   *Fl USSR*, 197; *Fl Iranica*, 112; *Fl Turkey*, 128; Grey, *Hardy Bulbs*.

A. carinatum (= keel of a boat)
Keeled Garlic
Linnaeus   1753

The membraneous bulbs are quite small, 1 cm (⅖ ins). The 30–60 cm (12–24 ins) stem carries sheathing, linear leaves with prominent veins. Most noticeable are the 2 unequal, tapering valves of the bract, one being very long, sometimes measuring 12 cm (4¾ ins). Loose flower heads carry unequal stalks, the outside ones curving downwards with purple flowers. The stamens are purple, longer than the tepals and dotted with yellow pollen (see illustration under *A. vineale*).

A. carinatum subsp. *carinatum*
(syn. *A. flexum*)

The flower head has bulbils and a variable number of flowers. It is a pest in the garden, the bulbils scattering far and wide. Even so the bloom on stems and flower heads, borne with undeniable grace, make this an elegant plant to admire—at a distance.

A. carinatum subsp. *pulchellum*
(syn. *A. flavum* var. *pulchellum*)

More often called plainly but inaccurately *A. pulchellum* (qv), this is an excellent garden plant. Europe, Asia Minor. HZ 3

SOURCES   *Fl Eur*, 71; *Fl USSR*, 103; *Fl Turkey*, 58; *Hortus III*. Illus: Fitter & Fitter, *Wild Flowers of Britain & N Europe*; Keble Martin, *Concise British Flora*; Polunin, *Fl Greece & Balkans*.

A. carolinianum
De Candolle (DC)   1804

The species name suggests that the plant comes from North America. De Candolle described this plant growing in Paris in 1804 believing that it had been introduced from Carolina. When it was found to be identical with *A. blandum* from the Himalayas there was no possibility of changing the name, because of the rules of precedence. So *A. carolinianum* it must remain.

A dense globular head, 2–3.5 cm (¾–1¼ ins), of pink flowers sits atop a stout stem, 20–40 cm (9–18 ins). Grey describes the plant as sweetly scented. Four–6 short, broad, curved leaves have a slightly glaucous tinge and are present at flowering. The flowers are cylindrical with pointed tepals much shorter than the stamens. The bulbs are over 3 cm (1 inch) long, with brown, scaly coats and form an article of diet for the Kirghis of Afghanistan.

An easy plant to grow from seed, amenable to pot culture and usually happy in a well-drained, sunny site. Flowering July–August on stony slopes around 3,000–4,500 m (9,000–13,000 ft), Afghanistan to central Nepal. HZ 2–3

SYNONYMS   *A. blandum*, *A. polyphyllum*, *A. obtusifolium*, *A. thomsonii*.

SOURCES   *Fl Iranica*, 11; Grey, *Hardy Bulbs*. Illus: Polunin, *Flowers of Himalaya*.

A. caspium
(Pallas) Bieberstein   1808

This *Allium*, with its magnificent flower head, is a native of sandy deserts in southeast Russia, Iran, Afghanistan and west Pakistan. The type plant comes from around the Caspian Sea. Seed has circulated in the exchanges as *A. caspicum*, a naming error.

The umbel is 5–20 cm (2–8 ins) in diameter, with unequal flower stalks 3–10 cm (1¼–4 ins), studded with bell-shaped flowers, a subdued lilac, tinged with green. Lilac stamens even longer than the tepals add to the airy beauty of the flower head. The 3–6 leaves are slightly shorter than the stumpy 10–35 cm (4–13 ins) stem; almost black tunics enclose the spherical bulbs.

A footnote in *Fl USSR* records that *A. baissunense* might be a very local, white-flowered form.

Flowering in early summer, this beauty will require an alpine house or bulb frame and summer dormancy. HZ 4

SYNONYMS   *Crinum caspium*, *A. brahuicum*. *A. baissunense*, *Amaryllis caspia*.

SOURCES   *Fl Eur*, 110; *Fl USSR*, 220: *Eur Garden Fl*, 80. Illus: *Bot Mag*, 4598 (1851); *Journal of the RHS*, April 1970, Part 4; *Fl Iranica*, 131.

A. cepa Onion
Linnaeus   1753

Everyone knows what an onion is, what it tastes like, its smell and doubtless how to grow it, usually in rows in the kitchen garden.

According to *Flora Europaea*, *A. cepa* belongs to Section Cepa having basically round bulbs, varying in size and shape according to the cultivar. Gigantic specimens turn up in every horticultural show. The outer covering hardly needs describing, membraneous, smooth and firm, in colour ranging between brown, red, yellow, green or white. The hollow stem may reach 100 cm (40 ins) and 3 cm (1¼ ins) in diameter, tapering towards the flower head from an inflated lower section. The hollow leaves are basal in their first year, later sheathing the lower one-sixth of the stem. A persistent spathe splits into 3 below the spherical umbel, 4–10 cm (1½–4 ins), which is packed with star-shaped flowers, each having green-striped, white tepals and exserted stamens. The stigma is simple, the seeds black and angular.

See also pp. 21–2.

Vegetable gardeners rarely see their onions flower, unless growing for seed. Incompetent growers who allow the odd onion to bolt will find the dried heads excellent for flower arranging.

Onions are usually grown as biennials, some cultivars produce umbels with bulbils and few flowers. Unknown in the wild, *A. cepa* is thought to be derived from the Central Asiatic *A. oschaninii*. Culture requires sunshine, good drainage and adequate doses of fertiliser to produce bulbs that ripen well and dry for storage. HZ 2

SOURCES *Fl Eur*, 20; *Fl USSR*, 92.

A. cernuum (cernuus = drooping)
Nodding Onion
Roth 1798

A. cernuum is probably the most widely grown of the American onions. As a garden plant in northern climates few alliums could be easier to cultivate, so undemanding and good natured. The elegant shape is distinctive, an onion odour is absent, the flowers stand up well to rain and can be cut for the house.

Very widely distributed across North America from Canada through to Mexico and across the continent, flourishing in moist soils in the mountains and in cool regions, it has very successfully survived the crossing of the Atlantic and Pacific oceans.

Identification is easy—not for nothing has it been called the Nodding Onion. Whereas other alliums may have flower buds that hang their heads subsequently straightening their stems with maturity, *A. cernuum* remains with a crooked stalk, hanging its individual flowers like small chandeliers.

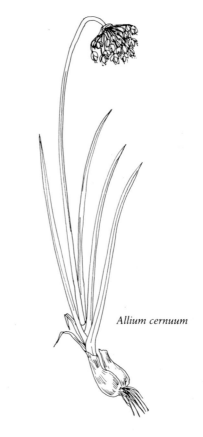

Allium cernuum

The narrow, greyish, membraneous bulbs are often clustered on a short rhizome. Under the outer covering the lining is pink or reddish. The stem, 10–50 cm (4–20 ins), may be cylindrical or flattened with slight winging, particularly towards the apex. The 1–6 leaves are shorter than the stem, 1–6 mm (¼ ins) broad and remain fresh and green at flowering time. Unlike many alliums, *A. cernuum* may produce 2 or more successive flowering stems from the same bulb. The umbel may carry 20–30 flower stalks above 2 bracts, the tepals varying in colour intensity, from pale pink to deep rose-purple with the whole measuring from 3–6.5 cm (1–2½ ins) in diameter. Cup-shaped flowers with tepals 4–6 mm (¼–⅓ ins) have long stamens, yellow or pink toned.

Once seed has formed the individual

flower stalks turn abruptly upwards from the point where they join the crooked stem, in a score of little S-bends, a reminder to gather the maturing heads if seedlings are not desired.

Two geographical variants have been described in the Rocky Mountains.

A. cernuum var. *obtusum*

Has red or pinkish bulb coats. This is the only form in the Rockies from Montana to New Mexico and Arizona according to Marion Ownbey in 'The Genus *Allium* in Idaho'. The stem is comparatively short with thicker, narrower leaves, curved on cross section.

A. cernum var. *neomexicanum*

Less stocky than the above, this plant is said to have keeled, flattened leaves, slimmer, taller stems and is frequently described as producing plants of intenser colour. However, Ownbey's descriptions in 'The Genus *Allium* in Texas' and 'The Genus *Allium* in Arizona' suggests there may be little difference between them.

These minutiae may puzzle gardeners who have been led to believe pale forms to be var. *obtusum* and deep forms var. *neomexicanum*. Moral—pick the colour forms desired. Most shades come reasonably true to seed, usually flowering in two seasons.

A. recurvatum Rydberg, a synonym, has been described from the northern Rocky Mountains, as well as *A. alleghaniense* Small, a deep rose form from the East and *A. oxyphilum* Wherry, white to pale pink.

Flowering June–August, a place in the open garden suits *A. cernuum*, some shade in dry gardens and adequate moisture. Rarely out of leaf, plants can be propagated by division and seed. HZ 2

SOURCES   Illus: Hitchcock *et al.*, *Vasc Pl Pac N W*; Clark, *Wildflowers of the Pacific Northwest*; Hay & Synge, *Dictionary Garden Plants*.

A. chamaemoly
(chamae = on the ground, creeping, lowly; moly = magic herb)
Linnaeus   1753

This must be one of the smallest onions, for the flowers sit almost stalkless on the rosette of basal leaves. The bulbs are 0.5–1 cm (⅕–⅖ ins), with pitted, thick, parchment-like coverings. The leaves, 2–5, linear, with small hairs along the edges and sometimes on the upper surface, usually lie flat along the ground. Star-shaped flowers with white petals and green or purplish veining may number 2–20. The spathe splits into 2–4 lobes which persist; the flower stalks are quite long, 5–10 mm (¼–½ ins). Overall the umbel measures 1.5–2.5 cm (½–1 ins) in width.

A. chamaemoly flowers from December to March on the sandy slopes and open spaces in the Mediterranean region of southern Europe and North Africa. HZ 5

Grey calls this onion an attractive small plant for a very sheltered position in the rock garden. However, blooming so early in the year, a bulb frame or alpine house would probably benefit both plant and grower. Seed is produced readily but seems loath to germinate; bulbs are available from at least one specialist supplier in Britain.

SOURCES   *Fl Eur*, 38; *Fl Turkey*, 18; Grey, *Hardy Bulbs*; Polunin, *Fl of Greece & Balkans*. Illus: *Bot Mag*, 1203 (1809); Redouté, *Liliacées*, 6: 325 (1812).

A. chinense
G. Don   1827

Known as *Rakkyo*, and not dissimilar to Chives, the plants are cultivated for their bulbs and are often found growing in China around the edges of fields of other crops. An important vegetable in Japan also, with hundreds of tons exported annually, according to *Hortus III* they are little cultivated in the USA.

Clustered, narrow, ovoid bulbs have a

A. *christophii* in East Sheen, London (John Fielding)

membraneous covering, from which the stems rise to 25−30 cm (10−12 ins). The hollow, evergreen leaves have 3−5 angles and, while they are as long as the stem, appear almost basal. Under the loose, hemispherical umbel the spathe splits into 2. About 20 flowers with light rose-purple, 4−5 mm (²⁄₅ in) long tepals and projecting stamens are carried on 1−1.5 cm (²⁄₅–³⁄₅ ins) stalks.

While plants originated in China they are also found growing in the Assam Hills, presumably introduced; flowering in the autumn, from September to November as the new foliage forms. HZ 3

SYNONYMS   *A. bakeri*, *A. exsertum*.

SOURCES Ohwi, *Fl Japan*, 8; *Eur Garden Fl*, 46; *Hortus III*; *Fl Rei Pop Sin*, 81; Stearn, '*Allium* & *Milula* in C & E Himalaya'; J.G. Baker, 'On the Alliums of India, China and Japan', 8.

A. christophii
Trautvetter 1884

Probably one of the best known, and certainly among the more spectacular and popular bulbous plants, *A. christophii* has suffered from name changing. The rules of nomenclature give precedence to the name given in the first description of a plant published after Linnaeus had brought method into the problems of establishing plant identity. Unfortunately this confuses the ordinary gardener if research demands that old and trusted labels have to be replaced. For many years this plant was known as *A. albopilosum* and can be found still under this name in bulb catalogues and on plant stands.

The round, grey, papery bulbs produce a rather short stem, 15−40 cm (6−16 ins), to support the large, globular flowering head, up to 20 cm diameter (8 ins) and reported to carry up to 80 star-shaped florets. These are a pale bluish-violet with a metallic sheen and narrow, sharply pointed tepals. While each individual flower is about 1–1.8 cm (⅖–⅗ ins) across, each is held on a long, stiff, flower stalk to 11 cm long (4½ ins). When the bulb first comes into flower, it may be hemispherical but by maturity the whole resembles some huge Christmas tree bauble.

The strap-shaped leaves are slightly glaucous and hairy on the undersurfaces but have withered by flowering time. The seed head dries beautifully, remaining crisp and attractive for 2–3 years with care, naturally very popular for flower arranging. In Britain at least, the bulbs are inexpensive and easily obtainable from popular commercial sources.

In the wild, plants grow on rocky slopes around 1,000–2,000 m (3,000–6,000 ft), from Soviet Central Asia to Iran, flowering in May/June. In the garden, they are quite easy to grow in hot, dry spots but tend to rot in areas of high rainfall unless given very good drainage. *A. christophii* makes a most attractive pot plant. The bulbs enjoy a summer rest, seed is plentifully produced but may require 4–5 years to produce flowering plants. HZ 5

SYNONYM   *A. albopilosum.*

SOURCES   *Fl USSR*, 192; *Eur Garden Fl*; *Hortus III*. Illus: Synge & Hay, *Dictionary Garden Plants*; Rix & Phillips, *Bulb Book*; most illustrated bulb catalogues; *RHS Journal*, April 1970, Part 4, seed head arrangement by Julia Clements.

A. chrysantherum
(chryso = golden)
Boissier & Reuter   1882

Almost globular bulbs, 4 cm (1½ ins), support 50–80 cm (20–32 ins) stems with 6–8 leaves verging from hair-like to broad, 5–30 mm (to 1¼ ins) wide. Hemispherical umbels are packed with flowers on 3 cm (1

¼ ins) long pedicels, the outer being yellowish or having a green tint, while the inner may be purple or almost the same colour as the outer rings. The tepals soon lose their shape and have almost disappeared by the time seed forms. The stamens may be purple or yellow and tend to be fleshy, while the ovary is green or black.

A. reflexum was described by Boissier as having yellowish-white tepals, those of *A. chrysantherum* as white. Species' colouring appears to be very variable.

Found in steep banks among scrub in terra rossa in the Middle East, around 800–2,150 m (2,500–6,250 ft), flowering May and June.

Plants are doubtfully in cultivation but the garish description suggests that *A. chrysantherum* might be fun to grow.

SYNONYMS   *A. reflexum, A. eginense.*

SOURCES   *Fl Turkey*, 125; *Fl Iranica*, 111.

A. chrysanthum
(= golden flower)
Regel   1875

Although described over a century ago and found by such eminent collectors as Przewalski, Rock, Wilson and Kingdon-Ward, it seems probable that it is not in cultivation in Great Britain. This is regrettable for the descriptions of its golden-yellow, creamy-white or cream flowers sound interesting. Herbarium specimens dry a light yellow that glistens.

Originally found in the mountains of South Gansu, (Kansu) West China, it has since been reported from Tibet. Flower height varies from 20–50 cm (8–20 ins). Grey suggests that should it become available, moist slopes would be a suitable environment.

A. phariense closely resembles *A. chrysanthum*, though the flowers have a slightly reddish tinge.

SOURCES Stearn, 'Allium & Milula in C & E Himalaya', 5 & 6; Fl Rei Pop Sin 74; Grey, Hardy Bulbs.

A. chrysonemum
(nema = thread, filament)
Stearn 1978

Chrys-, chryso- being Greek for golden, here is another Allium with yellow in its composition, albeit of a pale greenish tint in place of glitter. Rocky, stony soil in south-east Spain around 670–2,000 m (2,000–6,000 ft) is the home of A. chrysonemum, flowering during the summer months.

A 30–60 cm stem (12–24 ins) arises from ovoid bulbs and is clothed over its lower third by 3–4 leaves. These are 15 cm (6 ins) long, hairy, hollow threads that have withered by anthesis. The 2 valves of the spathe are unequal to 1.5 cm (3/5 ins) and persistent below the rather lax, many-flowered umbel. The whole is around 4 cm in diameter (1½ ins), with pedicels that nod until turning upright as the seed ripens. The cup-shaped florets have green mid-veins and slightly protruding stamens, also yellow, but occasionally with a reddish tinge in the upper parts.

A frame or alpine house will be required for cultivation. Seed or plants are not easy to acquire as A. chrysonemum has a reputation for being difficult to grow. HZ 4

SOURCES Fl Eur, 47; Eur Garden Fl, 45. Illus: Annales Musei Goulandris, 4: 150 (1978); Pastor & Valdes, Revision del genero Allium, 103 (1983).

A. cilicicum
Boissier 1846

According to Flora Turkey, 89, this is probably a high mountain form of A. scorodoprasum subsp. rotundum.

A. circinnatum
(circinata = rolled spirally downwards)
Sieber 1823

This is a very small plant with a short hairy stem, 5–18 cm (2–7 ins), topped with 2–5 tiny, star-shaped, white flowers striped with pink, in a shuttlecock-shaped umbel. The 2–5 leaves are very hairy, narrow and lie on the ground. They are also twisted into 5–10 spiral coils.

This strange little plant grows on dry, stony hillsides and is only found on Crete. It collects an asterisk in the Bulb Book as endangered or very rare in the wild. However, as alliums grow readily from seed, plants if obtainable would require cultivation in bulb frame or alpine house. March, April. HZ 5

The Bulletin of the AGS, Vol. 44, p. 239 carries a quotation from C. Grey-Wilson: 'exceedingly queer; a plant to be grown for amusement in sink or alpine house and surveyed with a cynical expression through a stern lorgnette.' A. circinnatum is unlikely to feature as Allium of the Year but there may always be room for oddities . . . of small stature.

While the most usual spelling appears to be A. circinnatum, doubtless from the original description by Sieber, A. circinatum is also used, a more logical form considering the name derivation.

SOURCES Fl Eur, 24; Grey-Wilson & Mathew, Bulbs; Polunin, Fl of Greece & Balkans. Illus: Rix & Phillips, Bulb Book.

A. condensatum
(condensed = numerous flowers on short pedicels)
Turczaninow 1855

Seed is available of this plant from the Far East, flowering in rocky areas from Mongolia, China to Japan. From the small plants in the greenhouse, I expect to grow bulbs on short rhizomes with a 30–80 cm scape (12–30 ins) with 4–7 slightly shorter, hollow leaves. The almost spherical umbel should carry many bell-shaped, yellow flowers with a greenish nerve and long stamens. Flowering time, July/September, probably in the open garden. HZ 3

SOURCES   *Fl USSR*, 67; *Fl Rei Pop Sin*, 64.

A. controversum
Synonym of *A. sativum* var. *ophioscorodon* (qv).
(ophio = pertaining to snakes; scorodon = garlic) Serpent Garlic

SYNONYMS   *A. controversum* Schrader, *A. ophioscorodon* Link, *A. sativum* subsp. *ophioscorodon* (Link) J. Holub.

Before flowering the stem loops the loop, the wide coils being illustrated in an old print of 1601 in Clusius, *Rariorum Plantarum Historia*. Jason Hill describes it thus: 'beak-shaped flower heads writhe upward in elaborate curves and coils, imitating a flamingo at its toilet and sometimes tying themselves in knots.' He goes on to comment: 'it is quite in character that this studied performance is a prelude to nothing in particular and that the flower-head, when it last appears, discloses only a few onion flowers and a bunch of bulbils.'

This flamboyant quotation enlivens the 'Notes on the Genus *Allium* in the Old World' that Professor W.T. Stearn contributed to *Herbertia* in 1944. For a more mundane description, see *A. sativum*. HZ 4

Plants of '*A. cowanii*' that I have grown seem to have the same athletic urge.

The *RHS Dictionary of Gardening* 1956 lists *A. controversum* as a synonym of *A. pyrenaicum*; *Hortus III* places it under *A. sativum*.

A. convallarioides
(= resembling *Convallaria*, Lily-of-the-Valley)
Grossheim   1924

A pleasing group of this *Allium* was shown by the Royal Botanic Gardens, Kew in the RHS Lily Group exhibit in 1986.

One cm wide bulbs (⅖ ins) with grey, papery coats, associated with yellowish bulblets, produce a 40–60 cm stem (16–24 ins) sheathed to the middle with smooth leaf sheaths. The 4–5 narrowly linear, channelled leaves are shorter than the stem. The spathe may be twice as long as the densely packed umbel, which is free of bulbils. Unequal pedicels, 2–5 times longer than the flowers, carry small bracteoles. Yellow anthers, topping the stamens, hardly emerge beyond the edges of the white tepals; the style is further exserted. The capsule itself is slightly longer than the flower.

Flowering in June/July, Iran, Iraq, through the Caucasus into Central Asia (*Flora USSR* mentions in slightly saline places), plants will require the protection of a frame or alpine house. HZ 4

F. Kollmann in *Flora Turkey* lists *A. convallarioides* with a question mark in the lists of synonyms of *A. myrianthum*.

SYNONYM   *A. pallens*.

SOURCES   *Fl USSR*, 110; *Fl Eur*, 57; *Fl Turkey*, 69.

A. coryi
M.E. Jones   1930

Texas has a Yellow Rose proclaimed in song, this is the State's Yellow Onion. Flowering in April and May on rocky plains and slopes on the mountains of western Texas, it is rarely seen in cultivation.

The stem, 30 cm (12 ins), carries a few-flowered head, 10–25, with 2–3 shorter leaves that are still green at flowering time. The florets have longish, unequal stalks, the tepals are chrome yellow, occasionally tinged red and fading with age. As the seed matures the tepals turn papery and rigid.

A. coryi is closely related to *A. drummondii.*

Few gardens can reproduce the summer drying period of western Texas. As seed becomes available, bulb frame and alpine house should offer the best milieu for successful cultivation. HZ 4

SOURCES Ownbey, *Genus Allium in Texas; Hortus III.* Illus: Niehaus & Ripper *Field Guide to S-W U S & Texas;* Venning, *Guide to Field Identification Wildflowers of North America.*

A. cowanii
Lindley 1823

This is a non-valid name for *A. neapolitanum;* however, bulbs are available for sale under this label. The name has also been used to suggest that *A. cowanii* is a selected superior form of *A. neapolitanum* (qv).

Confusion arose from plants exported to the New World then reintroduced and renamed. *A. neapolitanum* seemingly left Europe as an Italian and returned as a South American, the Peruvian connection being named Cowan.

A. crenulatum
(crenulate = having small teeth)
Wiegand 1899

Hurricane Ridge in the Olympic Mountains, Washington, an area of *krummholtz* (alpine scrub), at 1,150–1,700 m (3,500–5,000 ft), has good roads, ranger stations and thousands of tourists are able to visit the area with the minimum of exertion. *A. crenulatum* must be one of the most commonly sighted 'high' alliums. For nine

months of the year the area is covered with snow, providing dry conditions throughout winter, yet the average annual precipitation, including rain, is above 355 cm (140 ins), an indication of the problems besetting those who grow plants from high places.

Neat, of small stature, 8–15 cm (3–6 ins), with 1–2 flattened, sickle-shaped leaves, still just green at flowering, this is a pleasing small allium for alpine house culture. The stem is flattened, double-edged or winged, with the margins minutely sawtoothed (hence the name). The tepals are pale pink with deeper pink mid-ribs, becoming rather papery as seed forms. The anthers may be yellowish or even purplish. This small onion occurs in the Cascade Mountains, Washington to Vancouver Island, Canada and down into Oregon.

In July 1988, on Hurricane Ridge, plants were suffering from attack by a Rust, one of the few instances noticed of disease in *Allium* in the wild.

Gravelly slopes, April/July. Cultivation will require dry winter conditions, while during the growing season plants are accustomed to hot sunshine, sharp drainage and heavy rainfall. HZ 4

SYNONYMS *A. vancouverense, A. watsonii, A. cascadense.*

SOURCES *Hortus III;* Grey, *Hardy Bulbs.* Illus: Hitchcock *et al., Vasc Pl Pac N W;* Rickett, *Wildflowers U S,* Vol. 5.

A. crispum Crinkled Onion
Greene 1888

While *Hortus III* calls *A. crispum* a synonym for *A. peninsulare,* other authorities attach species status.

This small onion from California has grey-brown, roundish bulbs, 1 cm (²⁄5) long. The stem is 10–20 cm (4–8 ins), with 2 slightly shorter, narrow leaves. The umbel carries 8–25 flowers ranging in colour from reddish-purple to orchid pink. The 3 outer tepals are slightly spreading, the inner are distinctly crinkled, giving the plant its soubriquet. A spathe that has split into 2 valves, stamens shorter than the tepals with yellowish anthers and a stigma usually forked into three, help in identification.

Flowering from March to May in heavy soil on foothills below 800 m (2,500 ft), *A. crispum* will require protection from summer rain in a bulb frame or alpine house. HZ 4–5

SYNONYM   *A. peninsulare* var. *crispum*.

SOURCES   Munz, *California Flora*. Illus: Abrams, *Illus Fl Pac States*, 1, No. 40: Rix & Phillips, *Bulb Book*.

A. cupanii
Rafinesque   1810

The egg-shaped bulbs have netted coats from which rise 10–25 cm (4–10 ins) stems, sheathed by threadlike leaves. The umbel is almost cone-shaped with 5–15 flowers, wrapped around by the spathe as florists's flowers are sheathed by a paper funnel. Small tubular, bell-shaped florets are pinky-white with a slightly darker central vein. The stamens are shorter than the tepals, with pale yellow anthers.

On dry, rocky ground from Italy around the Mediterranean towards Bulgaria, this small allium can be found in bloom from May–October; in Lancashire, usually August in the greenhouse.

While not unduly exciting, *A. cupanii* is quite dainty and seems to have occupied a pot for several years in an unnoticed corner of my greenhouse, as a reminder of European alliums when Asiatic species are flowering in the damp northern end of summer. Bulb frame or alpine house is necessary, with a degree of summer rest, the plant showing signs of some dormancy in May–June. HZ 4

Allium cupanii

Several other alliums are comparable, some in the Section Scorodon of the *Flora Europaea*, viz. *AA. callimischon, parciflorum, rouyi* and *obtusiflorum. A. peroninianum* may be identical and there are subspecies listed: *A. c.* subsp. *cupanii, anatolicum* and *hirtovaginatum*.

SYNONYMS   *A. hirtovaginatum, A. pisidicum*.

SOURCES   *Fl Eur*, 49; *Fl Turkey*, 20; *Eur Garden Fl*, 48. Illus: Grey-Wilson & Mathew, *Bulbs*; *Belmontia*, 7:19 (1977).

A. cuthbertii
Striped Garlic, Cuthbert's Onion
Small   1903

Comparatively few of the American Onions are found in the eastern States, even fewer members of the genus appear to thrive in warm and humid regions. The pot-grown plants seen have been most attractive, reminiscent of *A. unifolium*.

Fibrous, matted bulbs produce 3 or more flat leaves, usually withered by flowering time, which are shorter than the 15–35 cm (6–14 ins) stem. The umbel is held erect and does not produce bulbils, the perianth segments being white or rarely pink on 2–2.5 cm (¾–1 ins) stalks. While the tepals are narrow and rather pointed, the stamens alternate in length. The capsule is crested.

A. cuthbertii may be found on sandhills, granitic outcrops, in open woods and scrub oak, from north-east Florida into eastern Alabama and central South Carolina, central and north-west North Carolina, flowering April–July.

Alpine house culture will probably be necessary in most areas, with frost protection required. HZ 7

SOURCES *Hortus III*; Small, *Manual of Southeastern Flora*, 7. Illus: Duncan & Foote, *Wildflowers of the Southeastern United States*.

A. cyaneum
Regel 1875

Found by the botanist Przewalski in the north-west Chinese Province of Gansu (Kansu) and, despite Reginald Farrer's derogatory remarks about alliums in general, circulated by him. In the Appendix to *The English Rock Garden*, Farrer mentions that *A. cyaneum* had been distributed by Potanin via Petrograd but had not been cultivated in England.

William Purdom was Farrer's companion on some of his journeys. In 1914, plant hunting in Gansu (Kansu) and Tibet, several alliums were listed by 'F' collector's numbers. Purdom collected F 321 from shallow, limestone ledges on Lotus Mountain. He returned the following February to retrieve plants, hacking them from 1 m (3 ft) of ice. F 321 was named *A. purdomii*

A. cyaneum (Author)

but was later to be considered merely a synonym of *A. cyaneum*.

The ovoid bulbs, with long necks and dark grey coats, cluster on a short rhizome. The rather slender stem, 10–45 cm (4–10 ins), is sheathed at the base by 4 or 5 linear leaves shorter than itself. Above the solitary spathe the umbel is packed with brilliant blue, star-shaped flowers on very slender green stalks, 4–10 mm (¼–⅖ ins). The blue stamens are considerably longer than the tepals, an important feature in identification. Confusion with *A. beesianum* and *A. sikkimense*, whose stamens are either as long as or shorter than the tepals, is prevalent.

A. cyaneum is probably the most commonly grown blue Allium, a delightful plant for the rock garden, usually 10–15 cm (4–6 ins). C.H. Grey thought it should be in every garden and wrote that he hardly knew of any plant to compare with it for brilliancy of colour. If it has a fault, the leaf tips tend to be drying by flowering time but so profuse are the flowers on an established clump, this is a small fault. Flowering in August–September, *A. cyaneum* is ideal for providing late summer colour.

Propagation is easy by division, seed being an alternative method. In north-west England I prefer to divide in spring; our soil is growing colder and wetter by the time *A. cyaneum* has reached the end of flowering. Plants prefer a moist, well-drained soil, growing well in peaty conditions and revelling in Cumbria despite the marauding slug population. Its native habitat is open grassland around 2,500 m (7,500 ft). (Pl. p. 152.) HZ 5

SYNONYM   *A. purdomii.*

SOURCES   *Eur Garden Fl*, 11; *Fl Rei Pop Sin*, 36; Farrer, *English Rock Garden*, Appendix. Illus: Grey, *Hardy Bulbs*; Rix & Phillips, *Bulb Book*; Moore in *Baileya*, Vol. 3, 1955, Fig. 54; Synge, *Collins' Guide to Bulbs.*

A. cyathophorum
(cyath = cup; -phorum = carrier)
Bureau & Franchet   1891

The cylindrical bulbs, coats splitting into parallel fibres, are clustered on a short

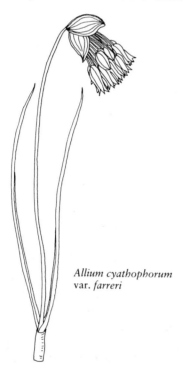

Allium cyathophorum var. farreri

rhizome. From this rises the angled stem, 20–40 cm (8–16 ins), surrounded by 3–6 narrowly linear leaves of similar length. The membraneous spathe may be single or splitting into 2 or 3 lobes, half as long as the flowers. 12–18 star-shaped, blue-violet flowers on 8–20 mm (⅓–¾ ins), unequal stalks, hang in one-sided clusters, becoming erect in fruit. The stamens, shorter than the tepals, have the lower halves of the filaments fused to form a tube.

This is the description of the type plant, found in the Chinese Province of Sichuan, flowering in June and doubtfully in cultivation.

Plants most commonly seen in cultivation are *A. cyathophorum* var. *farreri* (qv).

A. cyathophorum var. farreri
Stearn

Collected by Farrer and Purdom in the Province of Gansu (Kansu) and described by Stearn in 1930 as *A. farreri*, its correct status is now *A. cyathophorum* var. *farreri* (Stearn) Stearn. Further confusion occurs when plants are obtained by the gardener as *A.* sp. Tibet, or worse *A. tibeticum* (qv), the latter a valid species physically quite unlike the plant under discussion.

Several forms are found in gardens, some with erect stems and leaves, other with floppy leaves and drooping stems. The leaves are keeled at the base and channelled above, with a reddish sheath; the stem is slender and winged. There is also variation in plant size. Unlike the type, the flowers are reddish-purple and bell-shaped. Identification involves the one-sided pendulous umbel and the stamen tube. (Pl. p. 38.)

Forgetting botanical problems, *A. cyathophorum* var. *farreri* is a robust and decorative plant, ideal for gardens with acid soils and a high rainfall. Plants require moisture, albeit drainage, and will tolerate a remarkable degree of shade. In Lancashire, not noted for either summer heat or light intensity, one form of *A. cyathophorum* var. *farreri* flowers well in a raised bed that has over time become completely shaded by the outlying branches of a Rhododendron.

Seed is produced in abundance, requiring prompt dead heading, and explaining the monotony with which it appears in seed lists under a variety of pseudonyms. Recipients unaware of the aliases will at least get an excellent garden plant. Pendulous forms are lovely planted in a moist rock crevice and viewed from lower vantage points, an advantage in our precipitous Cumbrian garden.

Moist conditions or some shade and adequate soil depth are required if the leaves are not to dry at the tips during the flowering period. Propagation is easy by division, though seed presents no problems of fertility. Flowering May–July. HZ 2

SYNONYM   *A. farreri.*

SOURCES   *Eur Garden Fl,* 19; *Grey, Hardy Bulbs; Fl Rei Pop Sin,* 13. Illus: Rix & Philips, *Bulb Book;* Synge, *Collins' Guide to Bulbs;* Moore in *Baileya,* Vol. 2, 1954, Fig. 36.

A. cyrilli
Tenore   1828–1829

Membraneous ovoid bulbs, 1–2 cm (½–¾ ins) in diameter, produce 3–5 broadly linear, basal leaves, 20–30 cm (8–12 ins) × 1–3.5 cm (⅖–1¼ ins). Fifty–60 cm (20–24 ins) stems carry 4–7 cm (1½–2¾ ins) wide umbels, hemispherical or shuttlecock in form, packed with cup-shaped, white flowers striped with green. As seeds form the tepals curve backwards, with the tips reversing the curve and bending inwards, like carved hands on a Hindu temple. Included stamens bear yellowish anthers; the spathe splits into 3 lobes.

A. cyrilli is similar to *A. nigrum,* differing in the narrower tepals and their tendency to curve inwards at the tips.

Found on cultivated ground in southern Italy and southern and eastern Greece, *A. cyrilli* requires a warm, dry bed, bulb frame or pot culture, summer dormancy being necessary. April–June. HZ 4

SYNONYM   *A. nigrum* var. *cyrilli.*

SOURCES   *Fl Eur,* 107; *Fl Turkey,* 124; *Eur Garden Fl,* 66; Grey-Wilson & Mathew, *Bulbs.*

A. derderianum
Valek-e kouchek, Persian
Regel   1875

One of the small alliums with broad leaves, 1.5 cm (⅗ ins), and a short stem 4–10 cm (1½–4 ins), carrying a relatively large spherical umbel, 3–4 cm (1¼–1½ ins). The 2 basal leaves twist and curl to lie flat on the ground. Long whitish tepals, 8 mm (⅓ ins), with purple-violet nerves on unequal stalks appear extremely stiff and springy, remaining so after flowering.

Rather similar to *A. akaka* and *A. mirum, A. derderianum* makes an intriguing pot plant, requiring culture in a frame or alpine house. Found growing on dry slopes and steppe around 1,300–3,200 m (3,500–9,500 ft) in Iran and the Caucasus, flowering May. Nearer home, *A. derderianum* may be seen in RBG, Kew. HZ 5

SYNONYM   *A. haemanthoides* var. *lanceolatum.*

SOURCES   *Fl Iranica,* 95; *Fl USSR,* 188. Illus: Wendelbo, *Tulips and Irises of Iran.*

A. dichlamydeum
(chlamydia = a cloak)
Coastal Onion
Greene   1888

The ovoid bulbs have grey-brown coats with zigzag markings, the covering cracking on drying. One–3 flat leaves approxi-

A. dichlamydeum (Author)

A. d. var. *constrictum* and *A. d.* var. *nevii*, the last a smaller plant with narrower leaves.

The typical variety has ovoid bulbs with plain greyish-brown coats and purple, pink, red or white inner linings. A solid, round stem, 20–30 cm (8–12 ins), carries an umbel with a variable number of flowers and sheathed by 2 or 3 ribbed bracts. Pink flowers hang on slender stalks, which are longer than the narrow tepals, 7–8 mm (³⁄₁₀ ins), with prominent mid-ribs, and they continue to droop as the seeds form. Remaining green at maturity, the 2 leaves are shorter than the stem, rather sickle-shaped and flattened on cross section. Plants resemble the *A. senescens* group.

Discovered by David Douglas in 1826 in the Blue Mountains behind Walla Walla, Washington State, *A. douglasii* and its

mate the length of the rather stout solid stem, 10–30 cm (4–12 ins). Each umbel may carry 6–20 deep rose–purple, bell-shaped flowers, the outer tepals rather spreading while the inner stand erectly, 1–1.5 cm (³⁄₅ ins). The yellow-anthered stamens are two-thirds the tepal length. Tiny crests top the capsule.

A. dichlamydeum is quite distinctive with its deeply coloured, quite substantial flowers. Attractive as a pot plant, it has not proved hardy outdoors in the wet, cool climate of Cumbria. A bulb frame or alpine house would be safest, for those without hot, dry corners in the garden. HZ 4

SYNONYM *A. serratum* var. *dichlamydeum* Jones.

SOURCES *Eur Garden Fl*, 92; Munz, *California Flora*, 24. Illus: Abrams, *Illus Fl Pac States*; *AGS Bulletin*, Vol. 52, p. 371.

A. douglasii Douglas's Onion
Hooker 1838

Four varieties are described: *A. douglasii* var. *douglasii*, *A. d.* var. *columbianum*,

Allium douglasii var. *douglasii*

varieties are found on low hills, in shallow soil that is wet in winter, dry in summer; through eastern Washington and Oregon to Idaho. While individual plants are not spectacular, a myriad flowers covering the open spaces between the firs, the plains rolling below to a far horizon, leave a memory that is unforgettable.

The bulbs are said to be mild and sweet to eat. Cultivation in the open garden is easy. HZ 1–2

SYNONYMS A. hendersonii, A. nevii.

SOURCES Grey, *Hardy Bulbs*; *Hortus III*. Illus: Hitchcock *et al.*, *Vasc Pl Pac N W*; Abrams, *Illus Fl Pac States*, 1, No. 6.

A. drummondii Prairie Onion
Regel 1875

The name A. *mutabile* has also been misused for this species, but probably applies to A. *canadense*.

The finely reticulated bulbs are frequently found in a cluster, with 1–2 bulbs below the outer covering. There may be 3 or more leaves per bulb, each as long as the rounded stem, 10–20 cm (4–8 ins), channelled and flattened on cross section, remaining green at maturity. Two–3 spathe valves enclose the erect umbel which carries 10–25 bell-shaped flowers. The tepals may be pink, white, red, or occasionally greenish-yellow, 6–9 mm (¼+ ins) long with a slightly thickened mid-rib. While the equal pedicels are 1–3 times longer than the tepals, the stamens are shorter. After fruiting the flower head becomes papery and rigid.

A. *drummondii* is found throughout Texas, western Nebraska, New Mexico and northern Mexico, on plains, hills and prairies, particularly in limestone areas. Plants are conspicuously absent on pasture land, sheep and goats having found them greatly to their liking.

Rather remarkably, A. *drummondii* can produce up to 3 flowering stems per bulb in succession when conditions are favourable. This ability is shared by the closely related A. *coryi* and A. *runyonii*.

Plants appear in commerce under the synonym A. *nuttallii*. English-born, Thomas Nuttall, 1786–1859, explored the Missouri and Arkansas rivers. Later Professor of Natural History at Harvard, he was the model for 'Old Curious', a naturalist in Richard Henry Dana's novel *Two Years Before the Mast*.

A. *drummondii* flourishes in well-drained rock gardens. Alpine house or bulb frame will be required in cold, wet areas. HZ 4

SYNONYMS A. nuttallii, A. helleri.

SOURCES *European Garden Flora*, 87; Ownbey, 'Genus *Allium* in Texas'. Illus: Rickett, *Wildflowers U S*, Vol. 4; *AGS Bulletin*, Vol. 52, p. 39 (A. *nuttallii*).

A. elatum
Regel 1884

Synonym of A. *macleanii*. Originally described by Regel and for long grown under A. *elatum*, plants of this name were seen to be identical to A. *macleanii*. An account of the latter had been published one year earlier by Baker, so a change of name was necessary.

Further confusion arose from the similarity to A. *giganteum*, some of the differences being clarified by Wendelbo.

SOURCES *Fl USSR*, 216; *RHS Dictionary*; *Hortus III*. Illus: *RHS Lily Year Book*, 1967.

A. ericetorum
(ereike = heath or heather, Gr)
Thore 1803

Bulbs may be as thick as 1.5 cm (⅗ ins), oblong with outer coats breaking into parallel fibres, clustering on a short rhizome. Three–4 flat, linear leaves, almost basal,

A. ericetorum (Author)

only sheathe the lower end of the 10–40 cm (4–16 ins) stem. The spathe splits into 2 unequal valves, shorter than the flower head. This is hemispherical, 1–2.5 cm (⅖–1 ins) wide, packed with white or yellowish, cup-shaped flowers on equal stalks. The brown-tipped stamens may be twice the tepal length, and accentuate the plant's daintiness.

Found on stony hillsides, heathland, 300–1,600 m (900–4,800 ft) throughout south-west France, northern parts of the Iberian peninsula, into northern and central Italy, the Carpathians and Yugoslavia, flowering July–September.

Plants will enjoy a warm dry position in the garden, with sharp drainage. HZ 4

SYNONYM   *A. ochroleucum.*

SOURCES   *Fl Eur*, 8; *Eur Garden Fl*, 4. Illus: Grey-Wilson & Mathew, *Bulbs*.

A. falcifolium
Scythe-leaved Onion
Hooker & Arnold   1841

Plain brownish coats clothe the ovoid bulbs. The stem, 5–10 cm (2–4 ins), part of which is below ground level, is flattened with 2 distinct edges. Also flattened are the 2 sickle-shaped leaves, 4–9 mm (⅕–⅖ ins) wide and markedly longer than the stem. A rather crowded umbel carries 10–30 bell-shaped flowers below which the spathe splits into 2 long, wide valves. Six–12 mm flower stalks (½–¾ ins) support deep rose tepals, rather narrow but spreading at the tips. The stamens are half the tepal length. (Pl. p. 46.)

A. falcifolium is a distinctive little Onion, growing on heavy or rocky ground usually on serpentine outcrops, 150–210 m (500–700 ft), from the Siskiyou Mountains in Southern Oregon to Sonoma Co., California. While the habitat appears arid and sun-baked the ground below the buried stem and roots remains surprisingly damp. Some very deep purple forms occur, also greenish-white tinged with pink, March–July. Leaf growth begins in late autumn, the flower buds having formed by Christmas.

The bulbs are not hardy in our corner of Cumbria. Needing late summer rest, bulb frame and alpine house are recommended. *A. falcifolium* makes an attractive pot plant, the rather fleshy leaves being green at maturity. HZ 4

SYNONYMS   *A. falcifolium* var. *demissum*, *A. f.* var. *breweri*, *A. breweri.*

SOURCES   *Eur Garden Fl*, 88; Munz, *California Flora*, 14. Illus: Rickett, *Wildflowers U S*, Vol. 5; Abrams, *Illus Fl Pac States*, 11.

A. fibrillum (fibrilla = fibre)
Jones's Onion
M.E. Jones   1902

David Douglas discovered *A. fibrillum* in the Blue Mountains of Oregon. It can still be found in abundance on the summits there in moist, shallow soils and from the Wallowa Mountains of south-east Washington through to Idaho and Montana, at around 1,700–2,400 m (5,000–7,200 ft). Close relatives are *A. madidum* and *A. brandegei*.

Ovoid bulbs do not reproduce by bulblets like *A. madidum* but by splitting. Slender stems, 10 cm (4 ins), carry open, many-flowered umbels. The flower stalks are no longer than the white or pale rose tepals, 8–12 mm (⅓–½ ins), which become papery as the seeds ripen. Contrasting with the tepals are the yellowish or purplish anthers. Still green at maturity and rather sickle-shaped, the 2 leaves equal the length of the stem.

Flowering May–July, plants require some moisture during the growing period but protection from winter wet. HZ 4

SYNONYM *A. collinum.*

SOURCES Illus: Abrams, *Illus Fl Pac States*, 34; Hitchcock *et al.*, *Vasc Pl Pac N W*; Niehaus & Ripper, *Field Guide Pac States Wildflowers*.

A. fimbriatum Fringed Onion
S. Watson   1879

The red-brown coats of the ovoid bulbs have no netting on the surface. A rather stout, solid stem, 10–15 cm (4–6 ins), is sheathed for over half its length by an overtopping solitary leaf. Loose umbels carry 10–20 purply-pink, small bell-shaped flowers, the outer ones of which are nodding. Stamens and style are only half the length of the tepals. Disappointingly, the fringe of the title refers only to the crests on the ovaries.

There are numerous varieties described: *A. fimbriatum* vars. *purdyi, abramsii, sharsmithae, denticulatum, diabolense, mohavense, munzii* and *parryi*. *A. fimbriatum* and these regional variations are found at 700–2,500 m (2,000–8,000 ft) on dry slopes, very often in heavy soil, from the Colorado and Mohave Deserts to the Coast Ranges, California. Associated plants are Pinyon Pine and Creosote Bush. March–July.

Cultivation requires summer heat and aridity with winter protection; light sandy soil in warm gardens, bulb frame or alpine house. HZ 5

SOURCES Munz, *California Flora*, 31; Grey, *Hardy Bulbs*. Illus: Abrams, *Illus Fl Pac States*, 24; Munz, *Calif Desert Wild Flowers*; Niehaus & Ripper, *Field Guide Pac States Wildflowers*.

A. fistulosum
Welsh Onion, Japanese Bunching Onion
Linnaeus   1753

Cylindrical bulbs, 1–2.5 cm (⅖–1 ins), arise from a short rhizome. The hollow stem, 12–70 cm (4½–28 ins), has a swollen middle section which tapers to-

A. fistulosum in Denver Botanic Garden, Colorado (Author)

wards the base of the umbel. Two–6, hollow, cylindrical leaves, shorter than the stem, sheathe it for the lower one-third of its length. The 1–2 bracts are as long as the flower head, 1.5–5 cm diameter (³/5–2 ins), packed with yellowish-white, conically bell-shaped flowers on unequal stalks. The very much longer stamens are capped with yellow anthers. A pneumatic looking plant.

A. *fistulosum* is widely-grown as a vegetable, flowering in summer. HZ 1

Originating in the Far East, and known as Japanese Bunching Onion, it no longer occurs in the wild. The Chinese use the blanched bases of the stem. Perfectly hardy, the top growth appears early in the year and can be used as 'winter onion'.

The name Welsh Onion derives from *welische*—stranger. Vegetable gardening books are likely sources of illustration, A. *fistulosum* being infrequently grown as a decorative plant. A. *altaicum* is a related species and probably the wild ancestor.

SOURCES   *Fl USSR* (88); *Fl Eur*, 21; *Eur Garden Fl*, 26. Illus: Moore, *Baileya*, Vol. 3, Fig. 53.

A. *flavum*
Small Yellow Onion
Linnaeus   1753

One of the most commonly grown alliums, valued for its impeccable good manners as well as its colour. Varying in height, there are tiny forms suitable for troughs, others for the front of a border, with every size in between. Self-seeding is not a problem with the several forms I grow in north-west England.

Membraneous, ovoid bulbs, 1–1.5 cm (²/5–³/5 ins), support 2–3 leaves, 20 cm (8 ins), sheathing the lower third of the stem. This is 8–50 cm (3–20 ins), topped by a long spathe which narrows into an even longer slender appendage, 11 cm (4 ins). This spathe is a very prominent feature of A. *flavum*. As the buds form, the spathe stands erect, reminiscent of a head of barley. Then as the buds plump out, beginning to burst from their confinement, imagine a patient with mumps or a swollen jaw tied up in a long fanciful scarf that for some strange reason is standing on end. The effect is both comical and pleasing. As the flowers escape, the inner least mature ones continue standing erect while the first to open hang downwards. Meanwhile the spathe splits into 2 valves markedly unequal in length, which no longer point heavenwards but begin to droop until they tail away to earth as the seed forms.

The umbel may carry as many as 60 flowers and never produces bulbils. Unequal flower stalks 3–25 mm (¹/5–1 ins), carry the bell-shaped, lemon-yellow flowers, with the prominent exserted yellow stamens sometimes bearing lilac anthers. As seed forms the flower stalk once more begins to stand erectly.

A. *flavum* is found on dry slopes throughout southern and south Central Europe and into southern Russia, but oddly not on the Iberian Peninsula, according to *Flora Europaea*. June–July. HZ 3

In the garden some forms continue flowering into August. Well-drained borders and rock gardens will suit most plants; a summer rest is not necessary; plants grown in pots will languish if kept too dry. A. *flavum* is an excellent *Allium* species to recommend for northern gardens.

Several garden variants are obtainable. While the leaves and stem are normally glaucous, there are plants with very blue foliage, well worth seeking. This trait comes reasonably true from seed. Dwarf plants circulate under the labels of 'var. nanum', or 'ex- Wisley', non-valid names but common in horticulture. One attractive form attains a mere 6–8 cm (2½ ins).

A. f. subsp. *flavum*

Having yellow tepals and purple/yellow filaments, it appears in the usual range excluding USSR. Dwarf mountain variants are listed as *A. f.* var. *calabrum* and *A. f.* var. *minus*, and including synonyms, *A. quicciardii*, *A. nebrodense*, *A. callistemon* and *A. webbii*.

A. f. subsp. *tauricum*

Prevalent throughout south-east Europe from south-east Russia to Greece. May have tepals yellowish or nearly white, tinted with brown, pink or green. The filaments may be purplish towards the tips. This subspecies has the following synonyms: *A. flavum* var. *tauricum*, *A. paczoskianum* and *A. tauricum*. Possibly *A. sphaeropodum* may fit into this subspecies.

A. f. subsp. *tauricum* var. *pilosum*

Described by Kollman & Koyuncu, 1983.

A. flavum in Hartsop, Cumbria (Author)

A. flavum is very similar to *A. carinatum* and its subspecies *pulchellum*. Seed of plants with pink flowers known as *A. flavum pumilum roseum* appear in lists; the name is invalid and probably relates to dwarf forms of *A. carinatum* subsp. *pulchellum*.

SOURCES *Fl Eur*, 68; *Flora Turkey*, 55; *Eur Garden Fl*. Illus: Rix & Phillips, *Bulb Book*; Synge & Hay, *Dictionary Garden Plants*; Huxley, *Mountain Flowers in Colour*; McDonough in *Bull. ARGS*, p. 118.

A. forrestii
Diels 1912

While *A. forrestii* appears in the index to *Flora Reipublicae Popularis Sinicae*, if it is currently in cultivation in Great Britain, then it is very rare. This is strange, having been described as one of the best of George Forrest's (1973–1932) introductions from the Lichiang Range of western China.

With *A. beesianum* and *A. polyastrum*, *A. forrestii* was found growing above 4,000 m (12,000 ft) in moist grassland. With deep claret-red flowers, rather similar to *A. beesianum*, it should be an attractive plant for the garden, if re-introduced.

SOURCES *Bot Mag*, 9331 (not illus); Cowan, *G. Forrest, Journeys & Plant Introductions*, 1952; *AGS Bulletin*, Vol. 25; Blanshard, *JRHS*, April 70.

A. galanthum
Karelin & Kirilov 1842

Seed of *A. galanthum* frequently turns up on lists from Botanic Gardens. It closely resembles *A. cepa*, differing in having a solid stem and hollow leaves, while the middle of *A. cepa*'s hollow stem is inflated; there are also differences in the stamens. *A. galanthum* has 2–3 leaves against the other's 4–9. The stellate flowers are white, with stamens only slightly longer than the tepals.

Stony slopes in Siberia and Central Asia, flowering July. Culture in the open garden poses no difficulty. The bulbs are smaller than commercial onions, fully perennial and may be used in the kitchen similarly to the Welsh Onion, *A. fistulosum*. HZ 1

SYNONYM *A. pseudocepa*.

SOURCES *Fl USSR*, 89. Illus: *Bot Mag*,
1230; Moore in *Baileya*, Vol. 3, 1955,
p. 157.

A. geyeri Geyer's Onion
S. Watson 1879

Carl Andreas Geyer (1809–1853), an Aus-
trian botanist, travelled across the American continent in 1843 with a party of
missionaries. Throughout 1844–9 he continued to explore Washington State, Montana and Idaho, finally trekking down the
Columbia River. Sailing to England, his
collections were passed to Sir William
Hooker at Kew.

The *Allium* named after him grows in
Colorado, from north Mexico to Vancouver Island, from Texas to South
Dakota. Given its geographical range, it is
not surprising that synonyms abound.

The variant known as *A. pikeanum* was
found on Pike's Peak, Colorado around
1907 by Axel Rydberg and given this name
to commemorate the explorations of
Zebulon Montgomery Pike (1779–1813),
a US Army officer.

The ovoid bulb with its reticulated coat
has a long neck. Several grass-like leaves
surround the slender stem, 10–50 cm (4–
20 ins), and are still green at flowering
time. Ten–25 small, star-like, pale rose-
pink flowers on 2 cm (¾ ins) stalks become
rigid and spreading as the seed matures. All
the flowers of *A. geyeri* var. *geyeri* are
fertile, while there are bulbils in the head of
A. geyeri var. *tenerum*.

Despite its ubiquity, *A. geyeri* is not of
great garden merit, though the plant
known as *A. pikeanum* is small enough for
a trough. Found along streamsides and
meadows, the flowering period is May–
July. Some moisture is needed in the
flowering period but also protection in
winter. I have not found it hardy in
Cumbria. Frame or alpine house (Fig.
p. 16). HZ 3

SYNONYMS *A. fibrosum, A. rydbergii,
A. rubrum, A. arenicola, A. sabulicola,*
A. dictyotum, A. pikeanum, A. funi-
culosum.

SOURCES *Eur Garden Fl*, 84; Grey,
Hardy Bulbs; Hortus III. Illus: Hitchcock
et al., Vasc Pl Pac N W; Rickett, *Wildflowers U S*, Vols. 3 & 4; Anderson, *Wild
Flower Name Tales.*

A. giganteum
Regel 1883

Appropriately large, ovoid, leathery bulbs,
4–6 cm (1½–2½ ins), produce tall,
slightly ribbed stems, 80–200 cm (32–
80 ins). The large, floppy, glaucous, basal
leaves, half to one-third the length of the
scape and 5–10 cm (2–4 ins) wide, appear
above ground early in the year but have
mainly withered away by flowering time. A
single spathe encloses the large spherical
umbel, packed with star-shaped flowers.
Occasional white forms occur but the
usual colour is purple-violet. The flower
stems are almost equal, up to 6 cm (2¼
ins), while the tepals are elliptical and do
not change shape after flowering unlike
some of the other 'drumsticks'. The stamens are long and protruding, the ovary
warty on the surface.

A. giganteum, found in Iran, Afghanistan and Central Asia on lower mountain
slopes, requires a better-drained, warm
corner of the garden, than some of the
other 'drumsticks'. Bulbs are freely available commercially in the UK. April–May.
HZ 4

A. giganteum, spectacular in the garden, used for display in parks such as
Freeway in Seattle, does not dry well for
arrangements.

An excellent line drawing by Pat Davies
appears in the *Kew Magazine*, May 87,
accompanying R. Dadd's account of the
introduction of *A. giganteum*.

SYNONYM *A. procerum.*

SOURCES *Fl USSR*, 218; *RHS Dictionary; Hortus III*. Illus: *Fl Iranica*, 127; *Bot*

A. 'Globemaster' in the Royal Botanic Garden, Kew (Author)

Mag, 6828, 1885; Rix & Phillips, *Bulb Book.*

A. 'Globemaster'

Described as a hybrid between *A. stipitatum* and *A. giganteum*, this is a super-dramatic 'drumstick' at present not easily available. The magnificent heads may be seen flowering in the Royal Botanic Gardens, Kew. The leaves are large and reminiscent of *A. giganteum*. While the other parent is credited as *A. stipitatum*, the casual observer might wonder if a free-booter bee had visited *A. christophii* and smuggled in illicit pollen. Whatever the parentage, *A.* 'Globemaster' is a plant to covet.

A. glandulosum
Link & Otto 1828

When originally collected near Mexico City, the flower colour was described as deep red. Forms of this shade are still found in central Mexico. Further north the colour fades, only the mid-ribs remaining red. Other variants are described as maroon drying to purple. The glands of the title are drops of nectar from the ovary, relished by visiting insects.

The grey, ovoid bulbs have long slender rhizomes arising from the base, which

Ownbey likened to *Agropyron*, the ubiquitous Couch Grass. The 2–3 leaves are as long or longer than the 15–30 cm (6–12 ins), ridged stem. Remaining green at maturity, they are often toothed along the edges and nerves. The flower stalks are arching and unequal, 1.5–4 cm (⅗–1½ ins), twice as long as the tepals. Around 15 star-shaped flowers comprise the umbel, with stamens half the length of the tepals.

A. glandulosum inhabits meadows and moist slopes on mountains from west Texas to south-east Arizona and southwards into Mexico, flowering August to September and October. *A. kunthii* is similar, with smaller flowers, no rhizomes and inhabiting rocky outcrops, frequently of limestone. HZ 4

Considering the habitat, a warm rock garden with summer moisture should be the choice for cultivation, with protection for the late flowering season, alpine house or frame being a safer option. *A. glandulosum* is not common in cultivation. The unusual coloration suggests it would be an interesting addition to the autumn garden.

SYNONYM *A. rhizomatum.*

SOURCES Ownbey, 'Genus *Allium* in Texas', 'Genus *Allium* in Idaho'; *Eur Garden Fl,* 96. Illus: *Edward's Botanical Register,* 1826.

A. goodingii
Ownbey 1947

This new species was described by Dr Marion Ownbey in his paper on 'The Genus *Allium* in Arizona' published in *Herbertia,* 1947. Plants come easily from seed, flowering in their second year.

The bulbs terminate in a thick, Iris-like rhizome, from which arise several broad leaves, shorter than the stem, remaining green at anthesis. Eighteen to 23 flowers occupy the umbel atop a tall flattened

A. goodingii (Author)

SOURCES Ownbey, 'Genus *Allium* in Arizona'; Kearney *et al.*, *Arizona Flora*; McDonough, 'Allium Notes', *ARGS Bulletin*, Vol. 43, No. 1, 1985.

A. guttatum
(guttatus = spotted)
Steven 1809

A degree of uncertainty exists as to the correct naming of this species; a change to *A. margaritaceum* has been advocated.

One–2 cm diameter ovoid bulbs with yellowish bulblets form 10–90 cm (4–36 ins) stems, sheathed on the lower third by 2–5 shorter, threadlike or flattened leaves. This 1.5–5 cm (³/₅–2 ins) wide umbel is rather unusual, being packed with flowers on unequal stalks. The central ones, 3 cm (1½ ins), stand erect when forming seed, while the lower ones are very much shorter and stick outwards or point down suggesting an almost double flower head, while the whole base of the umbel is encircled by large membraneous bracteoles. The tepals vary in colour according to the subspecies, on a cylindrical flower with projecting yellow or purple stamens.

A. g. subsp. *dalmaticum*
(syn. *A. dalmaticum*)

Has tepals varying from pink to purple and is found from Albania, west Yugoslavia into Bulgaria.

A. g. subsp. *guttatum*
(syn. *A. margaritaceum* var. *guttatum*)

Has whitish tepals, keeled and blotched with purple, brown or green; grows from the Ukraine to the Aegean.

A. g. subsp. *sardoum*
(syns. *A. margaritaceum*,
A. sardoum, *A. gaditanum*,
A. confusum)

Lacks the blotch but the whitish segments

stem, 35–45 cm (14–18 ins). The flower stalks, 16–20 mm (½–¾ ins), curved and rather stout, carry elliptical pink tepals, withering by seed-time, with stamens of tepal-length.

A. goodingii belongs to a group containing *A. validum* from the Rocky Mountains over a wide North/South range, *A. brevistylum* from Montana to Colorado and *A. eurotophilum* found in northern Lower California. *A. goodingii* was found in the White Mountains of Arizona on steep rocky slopes around 3,000 m (9,500 ft), the original collection being made by Gooding in 1912. It has been confused with *A. plummerae*.

Seed sown in February 1986 flowered summer 1988 under dry conditions in the greenhouse in Lancashire, and has not yet been grown outside. If *A. goodingii* enjoys the same cultivation as the other similar *Allium* species then moist conditions will be required. HZ 2 carry a green or pinkish stripe; found from Portugal to the European areas of Turkey.

Found on dry slopes, roadsides and sandy beaches, flowering July/August. Frame, alpine house or warm dry corners of warm, dry gardens. HZ 5

SOURCES *Fl Eur*, 97; *Fl Iranica*, 106; *Fl USSR*, 155. Illus: Polunin, *Fl of Greece & Balkans*.

A. haematochiton
Red-skinned Onion
S. Watson 1879

I saw this most attractive *Allium* flowering early in March in Berkeley University Botanic Garden, California. One day I hope to grow it, albeit under glass, in Lancashire.

The long-necked, narrowly ovoid bulbs have deep reddish-purple, membraneous coats and usually grow in clusters. A 10–40 cm (4–16 ins) stem, accompanied by several flat, rather thick leaves the same length as itself, supports an umbel, 3–4 cm in diameter (1–1½ ins), of stellate flowers. Rose or dark purple tepals carry a deeper coloured mid-rib.

Growing on dry hillsides or beside streams, South Coast Ranges, California, from March–May, bulb frame or alpine house cultivation will be required. HZ 4

SYNONYM *A. marvinii.*

SOURCES *Hortus III*. Illus: Abrams, *Illus Fl Pac States*; Rickett, *Wildflowers of U S*, Vol. 4; Niehaus & Ripper, *Field Guide Pac States Wildflowers*.

A. heldreichii
Boissier 1859

The solitary bulbs are rounded with membraneous coverings from which rise 20–60 cm stems (8–24 ins). Two–4, shorter, hollow leaves sheathe the stem almost to one-third its length. The spathe splits into 2 persistent parts below the round or hemispherical head, packed with flowers on unequal 5–15 mm (⅕–⅗) stalks. Pink, bell-shaped with a deeper line down the middle, the flowers have yellowish stamens, shorter than the tepals.

A. heldreichii grows in rocky places, 700–2,000 m (2,000–6,000 ft), in northern Greece, June, July. HZ 5

Superficially looking like a rather leggy Chives, the bulbs are unlike the clustered, narrow bulbs of *A. schoenoprasum* and plants will require a drier, sunnier site with some degree of summer rest.

SOURCES *Fl Eur*, 104; *Eur Garden Fl*, 64; Boissier, *Fl Orientalis*, 14; *Hortus III*.

A. helicophyllum
(helic = coiled, spirally twisted)
Valak-e marpich—Persian
Vvedensky 1934

An intriguing little Onion from Iran, north Afghanistan and Turkestan, the short stumpy stem, 10–20 cm (4–8 ins), carries a loose, almost spherical head of star-shaped, brownish-violet flowers on unequal stalks. The most prominent features are the half-dozen leaves, 3 mm (⅕ in) broad and glaucous, the upper third twisted into corkscrew spirals. After fruiting the stem thickens and the pedicels lengthen.

Rare and not in general cultivation, this should be an interesting plant for pot culture. With increasing expeditions to Central Asia, hopefully plants will begin to be available.

SOURCES *Fl USSR*, 182; Blanchard, *JRHS*, April 1970. Illus: *Fl Iranica*, 132; Wendelbo, *Tulips and Irises of Iran*.

A. hirtifolium (hirti = hairy)
Boissier 1882

Roundish bulbs, whose outer covering cracks into papery strips, send up an 80–

120 cm (30–48 ins) stem, with a little ribbing. Each is topped with a bi-lobed spathe below the 'drumstick' head. Numerous star-shaped, linear, purple flowers, on 3.5–5 cm (1½–2 ins) long stalks, twist and reflex after anthesis. The stamens equal the tepals in length and the warty ovary sits on a short stalk. At ground level the 4–5 basal leaves have channelled lower parts, and are usually somewhat hairy. Occasional white forms occur.

Plants originate in Iraq, Iran and Turkey, flowering in May–June on rocky slopes and fields around, 1,500–2,500 m (4,500–7,500 ft). *A. stipitatum* is very similar.

A hot sunny border will suit *A. hirtifolium*. HZ 5

SYNONYM   *A. atropurpureum* var. *hirtulum.*

SOURCES   *Fl Turkey*, 140; *Fl Iranica*, 120; *Eur Garden Fl*, 73. Illus: Rix & Phillips, *Bulb Book* (showing plants with unribbed stems).

A. hyalinum
Paper-flowered Onion
Curran   1885

This Californian survived two seasons on an open raised bed in Cumbria before succumbing to the excessive rainfall; a high level of performance in such a damp climate. Shaded canyon slopes are the natural habitat, in the Sierra Nevada foothills, on moist rocky or grass slopes around 170–1,700 m (500–5,000 ft).

One cm (⅖ in) ovoid bulbs with grey, reticulated coats carry slender 15–30 cm (6–12 ins) stems. Two or 3 rather limp, flat leaves, as long as the stem, are still present at flowering, when the 6–10-flowered umbels carry white or pink-tinged, star-shaped florets. Widely spreading pedicels range between 2 and 2.5 cm (¾–1 ins), bearing tepals with a shiny patina that become almost transparent as

they age, explaining the species name. The umbel has a faint sweet fragrance, surprising those who imagine most alliums to be unpleasing to the nose. The stamens are white or cream, shorter than the tepals.

A. praecox is very similar in flower, differing in the detail of the bulb coat.

A. hyalinum is an attractive plant for the warm garden that can provide summer rest. Failing this, pot culture or a bulb frame is recommended. After flowering in spring the leaves wither, to re-emerge in autumn. A clear pale green with a marked mid-vein, they persist through the winter. HZ 3–4

SOURCES   Munz, *California Flora*; Grey, *Hardy Bulbs*. Illus: *AGS Bulletin*, Vol. 49, p. 325; Abrams, *Illus Fl Pac States*; Rickett, *Wildflowers U S*, Vol 4.

A. hymenorhizum
(hymen = membrane)
Ledebour   1830

A rather loose clump of bulbs, with thick, glossy, brown coats that split into vertical strips, attach to a rhizome. From these, stems 30–90 cm (12–36 ins), sheathed to half-way by 4–6 slightly shorter leaves, carry almost spherical umbels, 2–3.5 cm (¾–1¼ ins) in diameter. A dense mass of purplish-pink, bell-shaped flowers with tepals, 4–6 mm (¼ in) arise on nearly equal pedicels, 1–1.5 cm (⅖–⅗ ins) long. The projecting stamens have yellow anthers.

Flowering in damp meadows from July to August, from western Siberia, Central Asia, Afghanistan to Iran. *A. hymenorhizum* grows quite well in Lancashire in a dry sunny spot, albeit with reasonable drainage. The umbel is rather small in comparison to the length of stem for this to be a very significant garden plant. HZ 3

SYNONYM   *A. macrorhizum* Boissier.

SOURCES   *Fl USSR*, 56; *Fl Eur*, 13; *Eur Garden Fl*, 7. Illus: Ledebour, *Icones*, 4:

359 (1833); Moore in *Baileya*, Vol. 3, 1955, Fig. 53.

A. inderiense
Fischer 1838

A. inderiense resembles *A. insubricum* and *A. narcissiflorum* having slightly smaller, bell-shaped, purplish flowers with a deeper stripe. The beaked spathe differs by splitting into 2 lobes and the violet anthers are longer, reaching the tepal ends or even slightly protruding; the umbels also remain erect.

This is a plant of the Russian steppes, May–June. HZ 5

SYNONYMS *A. tataricum, A. diaphanum, A. beckerianum.*

SOURCES *Fl Eur*, 17; *Fl USSR*, 32; Grey-Wilson, *Bulbs.*

A. insubricum
Boissier & Reuter 1856

A plant invested with an aura of confusion. This would not matter if its flower were not one of the lovelier if not the loveliest of the whole genus. Plants appear in nurseries, in catalogues and in seed lists, far too often disguised as *A. narcissiflorum*. The latter is in fact rarely found in cultivation, despite the hundreds of labels that mark pots or garden beds.

Thin membranes surround the oblong bulbs, which grow in clusters on short rhizomes, leaving the bulbs with a curiously naked look when lifted from the ground. Stems may be 15–30 cm (6–12 ins), distinctly 2-edged, sheathed at their lower ends to the same level by 3–4 flat, glaucous, linear leaves still unwithered by flowering time. A single-valved spathe persists below the drooping umbel of 3–5 pinkish-wine-coloured bells with wide tepals, 6–9 mm × 18 mm (²⁄₅–³⁄₅ × ⁴⁄₅ ins) long. Paler pink forms also occur. The tepals flare outwards towards their edges,

A. insubricum (Author)

the bell transformed into the flowing skirt of an elegant ballgown, hiding the stamens and the 3-lobed stigma beneath its hem.

Found in calcareous, stony ground, 800–2,100 m (2,500–6,300 ft) in the Italian Alps, between Lake Como and Lake Garda, *A. insubricum* grows well in a raised gravel bed in Cumbria, 230 m (700 ft), with a rainfall of 305 cm (120 ins) per annum and acid soil. Plants grown in pots, with unheated greenhouse cover in winter, have fared less well though bulbs planted outside have also been lost in winters with unusually hard frosts. HZ 5

Seed germinates quite readily but flowering requires about three seasons. Offsets form but clumps of bulbs appear to dislike undue interference in our local climate. Propagation by removing outer bulbs without disturbing the main clump seems a safe procedure. (Similar stealth works well also for hellebores, *Ranunculus aconitifolius* 'Flore Pleno', and other plants that resent alteration in their settled way of life.)

After flowering, *A. insubricum* continues to nod its head while *A. narcissiflorum*'s seed heads straighten to stand erect. Further differing, *A. narcissiflorum* bulbs have layers of parallel fibres, the spathe may sometimes split into 2 or 3 lobes, also the flowers are narrower and more numerous, 5–8.

Particularly attractive in the rock garden, with persistent leaves flowering June, July, *A. insubricum* has a reputation for temperament. Possibly this relates to cultivation under unduly dry conditions. Summer 1988 brought 58 cm (23 ins) of rain through July and August to our fellside; in summer 1989 *A. insubricum* reappeared in greater numbers. Perhaps moist, well-drained soil without summer rest is quite acceptable. Meantime gardeners should remember where bulbs are planted, avoiding damage before the late-emerging leaves signal the plants' positions.

SOURCES *Fl Eur*, 16; *Eur Garden Fl*, 21. Illus: *Bot Mag*, 6182 (1875); cf. *A. narcissiflorum* in Rix & Phillips, *Bulb Book*.

A. jesdianum
Boissier & Buhse 1860

A. jesdianum is another tall 'drumstick' for the garden, with purple-lilac flowers, originating in Iran and Iraq.

Distinguished from the closely related *A. rosenbachianum* (qv) and *A. altissimum* (qv) by the netting on the covering to the bulbs, the flower stalks on the hemispherical umbel are slightly unequal and the stamens shorter than the tepals.

Flowering May–June, bulbs will require warm, dry conditions. HZ 4

SYNONYM *A. kazerouni.*

SOURCES *Eur Garden Fl*, 71; Grey, *Hardy Bulbs*; Mathew, *Large Bulbs*. Illus: *Fl Iranica*, 119.

A. karataviense
Regel 1875

Probably one of the best-known and grown dwarf *Allium* for the rock garden, part of its popularity arises from the use of its lovely dried heads for flower arranging. As a foliage plant alone, *A. karataviense* has great merit, with the broad grey-purple leaves appearing still fresh at flowering time. The tepals quickly dry after flowering, displaying the prominent ovary.

Easily grown from seed and relishing well-drained soil in full sun, *A. karataviense* is also effective as a pot plant. A native of Central Asia, from the Alai and western regions of the Tien Shan, in the wild, plants are found flowering in April and May in loose limestone scree. By June the plant has retreated into dormancy. HZ 4

Globular bulbs, 2–6 cm (¾–2½ ins), with blackish or greyish papery coats, throw up a dumpy stem, 10–25 cm (4–10 ins) in length. Frequently half of this is below soil level, the whole being shorter than the leaves. These are quite opulent, 3–15 cm (1¼–6 ins) broad and 15–23 cm

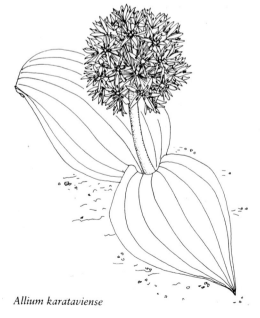

Allium karataviense

(6–9 ins) long, either grey-purple or dull green with a reddish margin and usually 2 in number. Two or 3 purplish-brown bracts surround the very large, dense, spherical umbel, 7.5–10 cm (3–4 ins), packed with pinkish-white or purple-tinged flowers. The flower stalks are rigid, stout, bright green, about 2.5 cm (1 ins) in length, carrying the star-shaped florets. The stamens are rather longer than the tepals, the white filaments carrying bright yellow, oblong anthers. The ovary is very noticeable. A white form is also in cultivation.

Illustrations may be found in many garden catalogues as bulbs can easily be obtained in commerce (Pl. p. 26).

SOURCES *Fl USSR*, 195; *Eur Garden Fl*, 77. Illus: Rix & Phillips, *Bulb Book*; Hay & Synge, *Dictionary Garden Plants*; Mathew, *Year Round Bulb Garden*.

A. lineare (Author)

A. lineare (Including *A. strictum* Schrader.)
Linnaeus 1753

Single or paired bulbs with brown, fibrous coats arise from a short rhizome and throw up a slightly ribbed, 25–60 cm (10–24 ins) stem, the lower third of which is sheathed by the shorter leaves. These are linear with tiny teeth on the edges and ribbed on the undersurfaces. Above the 2-valved, persistent spathe sits a dense, many-flowered umbel, 1.5–3 cm in diameter, (⅗–1¼ ins). Individual flowers are pink to purple-lilac bells with stamens that project to greater or lesser degree, while the style is strongly exserted. The capsule has broadly elliptic valves as long as the perianth segments.

Flowering in June–July on rocky banks, mountain slopes and grasslands from Central Europe through to the Ukraine, south and east Russia.

Plants have been quite hardy on a raised bed in Cumbria, and go quickly into seed. While the flower heads themselves are not exceptional, having a chunky solid quality,

the leaves are still green at anthesis. After the capsules split they dry to a pleasing creamy parchment. The seed heads will last for 3–4 years and are excellent for small dried arrangements. HZ 2

SOURCES *Fl Eur*, 11; *Fl USSR*, 13; Grey-Wilson & Mathew, *Bulbs*.

A. macleanii
Baker 1883

The plant long known as *A. elatum* has suffered a change of name, whereby the first published account takes precedence. Closely related to *A. giganteum*, *A. macleanii* is not so readily found in cultivation. Differences occur in the distinct ribbing of the stem which is 60–100 cm (24–40 ins), papery bulb coats, and a shorter spathe, one-third the width of the umbel. The leaves are a clear shining green, basal, shorter than the stem, 2–8 cm

(¾–3 ins) broad. Star-shaped, violet flowers, with slightly projecting stamens, do not reflex after flowering.

Found in Afghanistan, Central Asia, into south-west Asia on stony slopes in the upper mountain zone, plants were first collected by Colonel Maclean near Kabul.

Well-drained, sunny corners of the garden, flowering in June–July. HZ 4

SYNONYM   *A. elatum* Regel 1884.

SOURCES   *Fl Iranica*, 128; *Eur Garden Fl*, 76; Grey, *Hardy Bulbs*. Illus: *Bot Mag*, 6707, 1883; *RHS Lily Year Book*, 1967 (*A. elatum*).

A. macranthum
(large; -anthos = flower)
Baker 1874

Rhizomatous, oblong bulbs with membraneous coats throw up 3-angled stems, 20–45 cm (8–18 in), supporting a loose-flowered umbel carrying around 5–12 flowers, very occasionally up to 50. Rich plum-purple and bell-shaped, 8–12 mm (⅓–½ ins), with lemon, tepal-length stamens, they droop from 2.5–5 cm (1–2 ins) stalks. The leaves are channelled, 15–45 cm (6–18 ins), basal and rather flaccid; the spathe has disappeared before flowering.

While plants were collected in Sikkim in 1872 for Elwes, *A. macranthum* ranges from the Himalayas to south-west China. In the garden this is an excellent plant, good looking and easy. The flowers have a lovely grape-like bloom, on a 5 cm (2 ins) diameter head. Flowering late July–August, plants enjoy moist, well-drained soil, doing well in the open ground. Bulbs do not multiply rapidly but seed germinates well. HZ 4

SYNONYMS   *A. oviflorum, A. simethis.*

SOURCES   *Eur Garden Fl*, 15; Grey, *Hardy Bulbs*; Synge, *Collins' Guide to Bulbs*. Illus: *Bot Mag*, 6789 (1884).

A. macrum
(macrum = long, large or great)
Rock Onion
S. Watson 1879

'*A. douglasii* in miniature' has been written of *A. macrum*, replacing the former species in the Blue Mountains and drier eastern areas of Oregon and Washington. One can almost hear the crackle of barren gravel slopes in the evocative Kittitas and Klickitat sitings.

A small, brown, ovoid 1 cm (½ ins) bulb sends out 2 linear leaves, longer than the slender scape of 5–7 cm (2–3 ins). Slender flower stems form a spreading umbel containing whitish or pinkish flowers, veined with green or red-purple, becoming papery as the seed ripens. Stamens as long as the tepals carry reddish or purple-brown anthers, surrounding a grooved ovary.

Grey recommends a very attractive plant, easily grown in hot sandy soil, conditions not easily attained in north-west England. In cold, damp areas a bulb frame or alpine house will be necessary. April–May. HZ 4.

SYNONYM   *A. equicaeleste.*

SOURCES   Grey, *Hardy Bulbs*. Illus: Hitchcock *et al.*, *Vasc Pl Pac N W*; Abrams, *Illus Fl Pac States.*

A. mairei
H. Léveillé 1909

Edouard Maire collected specimens of this small onion in Yunnan in 1906. As Stearn comments, *A. mairei* was 'First described by Hector Léveillé in 1909, with a characteristically inadequate diagnosis'. Plants collected in Yunnan by George Forrest were later published by Diels in 1912, using the title *A. yunnanense*. Later work by Airy Shaw in 1931 on these two species added *A. pyrrhorrhizum*, another plant from the collections of Forrest. Two other similar *Allium* were *A. amabile* and *A.*

acidoides. While gardeners differentiate *A. amabile* and *A. mairei* by their colour, it has been suggested that members of the group are given varietal status under *A. mairei.*

The bulbs are very slender and the threadlike leaves, 10–25 cm (4–10 ins), sheathe the lower portion of the stem. 2–6 small, bell-shaped flowers 8–10 mm (1/3–1/2 ins) with pale to bright pink tepals (sometimes white with pink spots, sometimes pink-striped, the colour is variable), point upwards in a small wedge of an umbel. The stamens are shorter than the tepals.

In the garden, slugs are the main problem. If the clusters of small bulbs are allowed to attain a reasonable size before planting out, depredation does not wipe out the colony. *A. mairei* likes moisture, and grows well in the damp of north-west England, requiring however good drainage. Division in spring is safer than in autumn in cold gardens.

A. mairei and its co-species are attractive, small plants for the rock garden or raised bed, the leaves still fresh and green at anthesis. Self-seeding is not likely to be a problem. An added bonus is the late flowering season, August–early September. HZ 5

SOURCES Stearn, '*Allium* & *Milula* in C & E Himalaya', 8; *Eur Garden Fl*, 9; Mathew, *Dwarf Bulbs*; *Hortus III*. Illus: Rix & Phillips, *Bulb Book*, see *A. amabile.*

A. mirum (= wonderful, remarkable) Wendelbo 1958

Grey, papery coats enclose the 4 cm (1 1/2 ins) subglobose bulbs producing a thickish stem, 8–40 cm (3 1/4–16 ins). One–2 glaucous, wide rippled leaves, 1.5–8 cm (3/5–3 1/4 ins), purple-striped and slightly shorter than the scape, enhance the thickly packed, spherical umbel. Rather papery, campanulate flowers range from pale browny-purple to white with darker strip-

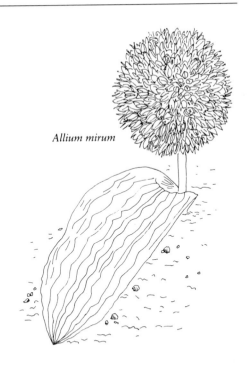

Allium mirum

ing on a 5–9 cm (2–3 1/2 ins) wide head. The long, exserted styles are an added attraction.

A. mirum has been compared to *A. karataviense*. The leaves have less blue in their colouring but the tennis ball head has better flowers. The deep shades fade as the flower ages but the texture is unchanged as seed forms. Among the collections made of *A. mirum* by Admiral Furse there were pure white forms.

A native of Afghanistan growing between 1,700–3,300 m (5,000–10,000 ft), flowering May, *A. mirum* is difficult to obtain, propagation relies mainly on seed, which is not yet available in the exchanges. Hopefully the situation will change in the next few years, for *A. mirum* is a delightful small plant that will be a great addition to alpine house collections. Plants may be seen in the Royal Botanic Gardens at Kew and Edinburgh.

Cultivation is not reputed to be too easy, hot dry conditions with excellent drainage being mandatory. HZ 5

SOURCES McDonough, *Bulletin ARGS*, Vol. 43, 2, 1985. Illus: *Fl Iranica*, 133; Mathew, *Smaller Bulbs, Dwarf Bulbs*; *RHS Lily Year Book*, 1967.

A. moly (= magic herb)
Linnaeus 1873

Rounded bulbs with parchment-like coverings produce grey-green, wide, basal leaves, 20–30 cm (10–12 ins) × 1.5–3.5 cm (³/₅–1½ ins), often appearing in twos and reminiscent of tulip foliage. Stems, 12–35 cm (4¾–14 ins) with 2-valved spathes, hold open umbels of bright yellow, starry flowers. The stamens are shorter than the tepals, the flower stalks unequal, 1.5–3.5 cm (³/₅–1¼ ins).

Probably one of the best-known onions, *A. moly* is a popular garden flower flourishing in the front of well-drained borders. Described by Parkinson as long ago as 1629, it was one of 16 species included in *Paradisi in Sole Paradisus Terrestris*. In sunny gardens, some shade may prevent the leaf tips scorching. Easily obtained from popular bulb catalogues, there are some forms that are shy of flowering, so that care in selecting vendors may be worthwhile. (Pl. p. 13, Fig. p. 48.)

A moly 'Jeannine'

This is a particularly fine form, with excellent foliage, firm, erect flowering stems and often producing 2 flower stalks from a single bulb. Originating in the Spanish Pyrenees, in 1978, and named after the wife of Michael Hoog, the Dutch nurseryman, this form was introduced by Brian Halliwell. Bulbs used as bedding can be seen in the Royal Botanic Gardens, Kew.

SOURCE *The Garden*, Dec. 1986.

A. moly var. *bulbilliferum*

Carrying bulbils in the head, this is not recommended unless as a curiosity.

Originating in east Spain and south-west France where it grows in shady, rocky areas, around 1,000–2,000 m (3,000–6,000 ft) flowering in early summer. HZ 3

SOURCES *Fl Eur*, 33; *Eur Garden Fl*, 35; Grey-Wilson & Mathew, *Bulbs*. Illus: Rix & Phillips, *Bulb Book*; *Bot Mag*, 499 (1800); Synge & Hay, *Dictionary Garden Plants*, 664.

A. moschatum (= musk-scented)
Linnaeus 1762

Slender stems, only 15–25 cm (6–10 ins), grow from narrow, ovoid bulbs covered by fibrous tunics, with threadlike leaves sheathing the bases. A dainty flower head, shuttlecock-shaped with quite long, almost equal pedicels, carries a few delicate flowers. These are bell-like, white or pink with a prominent dark vein on each segment. The tiny, dark violet anthers, like

A. moschatum (Author)

small commas, hang suspended inside the flower on almost invisible filaments.

Flowering between June–September, and found in rocky places and dry grassy areas throughout southern Europe from eastern Spain through to the Crimea and the Caucasus, below 2,000 m (6,000 ft). HZ 4

This is a delicate, fairy-like *Allium* for a pot in the alpine house or frame, flowering after the main flush of alliums is over. The fine leaves are still green at flowering and wither only as the seed is forming. A plant for growers who enjoy fragility in flowers.

SYNONYMS *A. capillare, A. setaceum.*

SOURCES *Fl Turkey*, 30; *Fl Eur*, 40; *Fl USSR*, 96; *Eur Garden Fl*, 43. Illus: Grey-Wilson & Mathew, *Bulbs.*

Several *Allium* resemble *A moschatum.*

A. inequale

Has very unequal pedicels. 3–6 times as long as the tepals. HZ 4

SOURCES *Fl Eur*, 41; *Fl USSR*, 95.

A. rubellum

Has yellow or brownish bulblets, hollow, threadlike leaves, many flowers in a dense, hemispherical umbel. These are bell-shaped, pink with a darker nerve but with yellow anthers. South-east Russia, Afghanistan, west Himalaya, where they grow on dry grasslands and semi-desert. May. HZ 4

SOURCES *Fl Eur*, 43; *Fl USSR*, 114; *Eur Garden Fl*, 44. Illus: Wendelbo, *Tulips and Irises of Iran.*

A. meteoricum

Has membraneous bulbs, a spathe with unequal valves, narrowing into a slender tail and unequal pedicels. The pink flowers are cylindrical, the anthers yellow. Rocky areas in the Balkans. HZ 4

SOURCES *Fl Eur*, 44; Key in Stearn, 'Genus *Allium* in the Balkan Peninsula'.

A. bornmuelleri

Has fibrous bulbs, tiny cilia on the leaf edges, spathe valves with tails, a loose, few-flowered umbel with pink, cup-shaped flowers lacking prominent veining but with purple anthers. Peloponnese, Greece, Yugoslavia. HZ 4

SOURCES *Fl Eur*, 42; Stearn, 'Genus *Allium* in the Balkan Peninsula'.

Others in this similar group are AA. *delicatulum, frigidum* (syn. *achaium*), *chrysonemum* (qv) and *grosii*. The last, from the Balearics, is a rare species with larger, darker purple flowers.

SOURCES *Fl Eur*, 45, 46, 47, 48.

A. murrayanum
Regel   1873

A degree of confusion surrounds the identity of *A. murrayanum*. Since the original description by Regel, *A. murrayanum* has been treated as a synonym of *A. acuminatum*. The plant found in gardens under this label, however, resembles *A. unifolium* and is almost certainly a variant of it.

The late Dr Marion Ownbey, authority on the American alliums, does not mention *A. murrayanum* in the entries under *A. acuminatum* that he wrote for *Vascular Plants of the Pacific North West* (1969), nor in his papers on the 'Genus *Allium* in Idaho' (1950) and the 'Genus *Allium* in Arizona' (1947). *A. cuspidatum* only is included as synonymic with *A. acuminatum*. H.E. Moore, listing 'The Cultivated Alliums' in the USA in 1955, has no mention, nor has *Hortus III* (1976); H.W. Rickett likewise in *Wildflowers of the United States.*

From British references, *The Botanical Magazine* has no appropriate entry in the

Index, Vol. 1–164, despite *A. mur-rayanum* being an attractive garden plant. C.H. Grey, *Hardy Bulbs* (1938) allows one line: 'Not more than a good form of *A. acuminatum*.'

RHS Dictionary (1956) is guarded: 'Nearly allied to *A. acuminatum*. Scape about 12 ins *fl* linear, longer than stem. *fl* rose-purple in large umbels. North America (G F 770).'

Meanwhile North America is a large place and curiously reticent about the species. The only references I have found both date from 1940, Morton, *A Checklist of Amaryllidaceae, Tribe Alliae in the United States*, in which he acknowledges the help received from Abrams, *Illustrated Flora of the Pacific States* (Vol. 1). In both cases the entries are merely as a synonym of *A. acuminatum* without comment and may stem from a single source. Both pre-date the aforementioned works of Dr Ownbey.

Printed references since 1938 are mainly found in British works, i.e. Synge, *Collins' Guide to Bulbs* (1961) where the illustration is more referrable to a form of *A. unifolium* with no sign of the tapering tepals diagnostic of *A. acuminatum*.

A taxonomic study would be of great interest to others who have also expressed doubts, including McDonough in *Bulletin ARGS*, Vol. 42, No. 4 (1984). In these circumstances, a description is not appended.

A. narcissiflorum
Villars 1779

Most of the plants available to gardeners, whether from commerce or from Society seed lists, will prove to be *A. insubricum* (qv). Differences between the two include several characteristics, most importantly and easiest to observe is the habit of *A. narcissiflorum* of producing a nodding flower head which grows erect as the umbel matures and seeds form. *A. insubricum* heads remain nodding until winter wind and rain destroy all vestiges of summer blooming.

The bulbs of *A. narcissiflorum* are coated with parallel fibres, umbels carry more numerous flowers, 5–8 wth narrower tepals, the spathe may split into 2–3 lobes and the flat, 2-sided leaves may be a darker green. (Pl. title page.)

The wine-purple flowers are found on stony ground in the south-west Alps, on surrounding hills in southern France and northern Italy and at one time reported in north-west Portugal, around 800–2,300 m, between July and August. In cultivation, the flowers are commonly pale pink.

Recommended as a good plant for the rock garden, all the plants that I have grown from seed or bulbs have turned out to be *A. insubricum*, flourishing in my wet Lakeland garden. *A. narcissiflorum* requires drier conditions.

Reginald Farrer devoted some resounding prose in favour of *A. narcissiflorum*. Hardly an *Allium* devotee, he claimed for this species the title of 'the glory of its race . . . in all the ranges of the world' not just 'in the steep earth-pans and stony screes high up in the most awesome shelves of the limestone Alps of Piedmont'. Sadly imagination then runs riot.

'Unfortunately an evil god-mother has dowered this beauty with a commensurate drawback, in the form of an exaggerated stench—a stench so horrible that one can hardly bear to collect it, to say nothing of the fact that its soil is like rock and one's own foothold slithering and insecure upon the lip of abysmal precipices.'

Poor *A. narcissiflorum*. Could Farrer's slithering feet, grinding plants to pulp, have caused the miasma? Few *Allium* offend the nose unless roughly handled.

Speculation has hovered around the specific name, for narcissus-like the flowers are not. An anonymous contributor to the Bulletin of the *AGS*, Vol. 53, p. 215 wrote that Villar used the name *A. narcissifolium*, a feasible description, as well as *A. narcissiflorum*. The same writer, assuming that *narcissiflorum* translated as beautiful flower, may have remembered

Narcissus, that handsome youth, obsessively hanging his head over a pool till his watery mirror reflected only a flower.

Many of the illustrations listed in books may well be *A. insubricum* rather than *A. narcissiflorum.*

SYNONYMS *A. grandiflorum, A. pedemontanum.*

SOURCES *Fl Eur*, 15; *Eur Garden Fl*, 20; Farrer, *English Rock Garden.* Illus: Grey-Wilson & Matthew, *Bulbs*; Synge & Hay, *Dictionary Garden Plants*; Synge, *Collins' Guide to Bulbs.*

A. neapolitanum Naples Garlic
Cyrillo   1788

Round bulbs produce 3-angled stems, 20–50 cm (12–20 ins), sheathed on their lower quarter by linear leaves with narrow keels on the undersurfaces, that while hairless may have minute teeth along the margins. Below the many-flowered, shuttlecock or hemispherical umbel sits a single, lobed spathe. The nodding buds on long 1.5–3.5 cm (3/5–1 1/2 ins) stalks open into erectly held large, star-shaped flowers with glistening white tepals, 7–12 mm (2/5–1/2 ins), longer than the yellow anthers.

Naples Garlic, found in dry, open places around the Mediterranean and Portugal, in spring, has been used as a cut flower by florists, forced, dyed with inks and even sprayed with paint, but is still a pretty plant to grow in warm gardens in sunny corners with good drainage. Plants survive in beds around the house in Cumbria under the rather deep eaves but require sun to produce flowers. This has been a surprise, for *A. neapolitanum* is considered slightly tender, requiring too a period of summer rest. The flowers are pleasant smelling, bulbils are not formed and certainly in north-west England self-seeding has not been the problem that can occur in warmer areas. *A. neapolitanum* makes an attractive pot plant if not allowed to grow too straggly. HZ 4

A. neapolitanum (Author)

Plants are easily available commercially, seed germinates well so that is a simple matter to grow sufficient plants for cutting.

A. cowanii

This appears in bulb lists, a form with large flowers, but is considered a synonym of A. neapolitanum.

A. 'The Pearl'

Has dark anthers and appears to be a hybrid.

SYNONYMS A. cowanii, A. album, A. lacteum, A. sulcatum, A. sieberianum.

SOURCES Fl Eur, 28; Eur Garden Fl, 31; Fl Turkey, 12. Illus: Rix & Phillips, Bulb Book; Grey-Wilson & Mathew, Bulbs; Synge, Collins' Guide to Bulbs.

A. nerinifolium
Herbert 1844

Synonym of Caloscordum nerinifolium. This plant, listed as A. nerinifolium Baker in Flora USSR, 226, is similar to Allium

A. 'The Pearl' (Author)

but has the stamen filaments fused to the tepals. Now removed to Caloscordum, plants are native to Mongolia, China and eastern Siberia, flowering on dry slopes in July–August.

SOURCES Fl Rei Pop Sin, 99 as A. neriniflorum. Illus: Rix & Phillips, Bulb Book.

A. nigrum
Linnaeus 1762

Membraneous, ovoid bulbs, 2.5–3 cm (1–1¼ ins) send up a thick, 60–90 cm (24–36 ins) stem, with 3–6 basal, broadly linear leaves, up to 8 cm wide and 50 cm long (3¼ × 20 ins). The single spathe becomes 3–4-lobed, below the hemispherical or slightly fastigiate umbel. Star-shaped, white or palish lilac tepals with greenish mid-veins on long stalks 2.5–4.5 cm (1–1¾ ins), make up the multi-flowered head; occasionally there are bulbils present. The stamens are shorter than the tepals and fused at their bases to form a ring, topped with yellow anthers. The most prominent features are the greenish-black, lobed ovaries, which give the species its name.

Syn. A. multibulbosum

Applies to forms with multiple bulbils.

A. auctum

Very similar; probably not a distinct species.

A. decipiens

Resembles A. nigrum but the leaves are much narrower, only 2–10 mm wide.

A native of southern Europe and Asia Minor, western Asia and North Africa, flowering from April–June in fields and calcareous rocks, 100–2,000 m (300–6,000 ft). Good forms are quite decorative. A. nigrum grows well in warm, dry gardens providing summer rest, otherwise bulb frame or pot culture. HZ 4

SYNONYMS *A. multibulbosum*, *A. monspessulanum*, *A. magicum.*

SOURCES *Fl Eur*, 106; *Fl Turkey*, 122; *Eur Garden Fl*, 65; Grey, *Hardy Bulbs.* Illus: Grey-Wilson & Mathew, *Bulbs*; Rix & Philips, *Bulb Book*; *Bot Mag*, 1148 (*A. multibulbosum*).

A. noëanum
Reuter 1875
Valak- e souri, Persian

A distinctive small onion, to 30 cm (12 ins), with 3–4 broad leaves standing upright from a basal rosette. The funnel-shaped umbel is held like an ice cream in a cornet by the purple-flushed spathe. After anthesis the seed head opens out to become hemispherical or even spherical. Sweetly scented, narrow bell-shaped flowers, pale or lilac-pink to deep mauve, fill the umbel on long, unequal stalks, 3.5–7 cm (1¼–2¾ ins). Purple anthers top red or blackish-purple filaments.

Flowering April–May in rocky ground, heavy clay and cultivated land, mainly in south-east Anatolia, Iran, Syria and Iraq.

A bulb frame with summer baking will be necessary when plants become available. HZ 5

SYNONYMS *A. jenischianum*, *A. dilutum.*

SOURCES *Fl Iranica*, 106; *Fl Turkey*, 135; Mathew, *Dwarf Bulbs*. Illus: Wendelbo, *Tulips and Irises of Iran*.

A. nutans
Linnaeus 1753

Conical bulbs are found attached to a stout rhizome, their tunics blackish and membraneous. The stem too is robust, 20–60 cm (8–24 ins), often winged at the upper end, carrying a spherical head, 5 cm (2 in) diameter, which as the name suggests is drooping before flowering. The equal pedicels have small bracteoles and are almost twice the length of the tepals. These are rose with a tinge of violet, the inner segments longer than the outer, the stamens almost twice as long again. Six–8 glaucous, basal leaves, unkeeled, half as long as the scape and slightly sickle-shaped, are still fresh when the flowers open.

Native to the steppes, stony slopes and meadows of Central Asia to Siberia, bulbs are easily grown in a sunny border, making attractive garden plants for June–July. HZ 1

SOURCES *Fl USSR*, 45; *Eur Garden Fl*, 2; *Hortus III*. Illus: Rix & Phillips, *Bulb Book*; *Bot Mag*, 1808, t.1143.

A. obliquum
Linnaeus 1753

A reddish-brown tunic encloses the thickish, oblong, solitary bulb attached to a horizontal rhizome. Rather long and lanky stems, 60–100 cm (24–40 ins), are sheathed half-way by a series of leaves, 6–10 in number. These are linear, channelled, 35 cm × 2 cm (14 × ¾ ins) and usually pointing upwards. The spathe splits into 2–3 lobes below the umbel, 3.5–4 cm (1½ ins), which is usually spherical or almost so, packed with pale yellow, cup-shaped flowers on nearly equal stalks, 1–2 cm (²⁄₅–¾ ins). Both the style and the stamens, tipped with slightly darker anthers, are longer than the markedly concave tepals. (Pl. p. 24.)

C.H. Grey in his book *Hardy Bulbs* is damning, writing of 'small greenish-yellow, evil-smelling flowers in May', a statement with which I totally disagree. The pale yellow flowers blend very happily in a mixed border in the Lancashire garden, a soft shade more agreeable than many strident yellows that accompany high summer. Perfectly happy in heavy clay and a high rainfall. *A. obliquum* flowers obligingly in June and July. Individual flowers are not outstanding but a sturdy clump is quite pleasing. There is no obvious onion smell.

A native of Central Asia and south-east Russia, where it blooms in meadows and wooded slopes, cultivation in the open garden should not cause problems. HZ 3

SOURCES   *Fl Eur*, 14; *Fl USSR*, 53. Illus: Rix & Phillips, *Bulb Book*; *Bot Mag*, 1408 (1811).

A. oleraceum (oleraceous = pertaining to kitchen gardens (as crop or weed)) Field Garlic
Linnaeus   1753

This is one of the plants that give onions a bad name. Any umbel carrying bulbils should be viewed with grave suspicion. Despite earning an entry in the *European Garden Flora*, *A. oleraceum* should be avoided. The smaller the bulbil the more

Allium oleraceum

likely it is to evade the gardener's fingers, rolling away to start another generation. Other paid-up members of this gang are *AA. vineale, carinatum* and *scorodoprasum*.

Arising from a small bulb, the stem, 25–100 cm (10–40 ins), is sheathed for half its length by 2–4 linear or threadlike leaves which are hollow in their lower ends. The spathe splits into 2 unequal valves, the longer growing to 20 cm (8 ins). Open umbels may carry as many as 40 flowers, a mixture of flowers and bulbils or be composed entirely of bulbils. What flowers there are may be bell-like, whitish with tinges of green, brown or pink, on unequal pedicels, the outer of which hang downwards. The stamens are shorter than the pedicels, tipped with yellow or reddish anthers.

A. oleraceum can be found on roadsides, scrubland, cultivated land or rocky slopes throughout most of Europe, into the Caucasus. Keble Martin records it as being widely distributed in Great Britain but rare. July–September. HZ 1

The flowers can be quite pretty and resemble *A. paniculatum*, but the bulbils should never be underestimated.

Cultural details are probably superfluous.

SYNONYM   *A. scabrum*.

SOURCES   *Fl Eur*, 63; *Fl USSR*, 104; *Eur Garden Fl* 54. Illus: Grey-Wilson & Mathew, *Bulbs*; Keble Martin, *Concise British Flora*, Pl. 85.

A. olympicum
Boissier   1844

Oval bulbs with blackish tunics, and 30–50 cm (12–20 ins) stems support dense, many-flowered, hemispherical umbels 2–3 cm (¾–1¼ ins) in diameter. Below, the awl-shaped valves are longer than the umbel. Violet stamens and a long style project from the violet-red or lilac-pink flowers growing on slightly unequal pedicels, 1–2 cm (²⁄₅–¾ ins), and more than twice the tepal length. There are 2–3 linear, flat leaves.

Found at 1,300–2,800 m (4,000–8,500 ft) above the tree line in scrub or coniferous forests and alpine meadows, northern Anatolia, Greece, flowering July–August. Dwarfer plants are also reported from higher altitudes.

A. olympicum has many features in common with *A. carinatum* ssp. *pulchellum* and *A. stamineum*. Some excellent plants for the rock garden have circulated under this species' name which do not quite correspond to the diagnosis of *A. olympicum*. One is described below under *A.* sp. aff. *stamineum* (qv).

Several *Allium* with collectors' numbers were listed as *A. olympicum* in the *Bulletin of the AGS*, Vol. 39, pp. 297, 303: AC&W 1956, AC&W 2353 and 2354 from the collecting tour of Turkey in 1966 of Albury, Cheese and Watson.

Bulbs from the catalogue of Paul Christian were described thus: twisted, narrow, cylindrical leaves and a golfball-sized spike of white-pink flowers borne close to the ground in July on a very compact plant. Stock from AC&W 2375.

Yet another *A. olympicum*, AC&W 2372, is described by Mark McDonough in *ARGS Bulletin*, Vol. 43, p. 1, with an excellent line drawing which correlates with the entry *A.* sp. aff. *stamineum*.

A description in *AGS Bulletin*, Vol. 45, p. 343 by V. Horton of *A. olympicum* but without a specific AC&W number is closer to the plant entry *A.* sp. aff. *stamineum*.

SOURCE *Fl Turkey*, 60.

A. sp. aff. *stamineum*

One of the most delightful alliums I have grown came from the AGS Seed List as *A. olympicum*. Flowering a year and a half after sowing it has flourished on a raised bed in Cumbria, proving its ability to survive heavy rainfall. The bulbs are increas-

A. olympicum, of gardens, in Hartsop, Cumbria (Author)

ing very slowly and seed production has been poor, but the plants have flowered yearly without any winter protection.

Four–6 glaucous, fistulous leaves sheathe the base of the stem. While the lower ones have withered by anthesis, the upper 1–2 remain fresh, growing longer than the scape but arching downward. A markedly blue-tinged scape, 5–10 cm (2–4 ins) in length, prostrates itself to half its length along the ground before lifting the umbel into the air. Pretty, loose, hemispherical heads of rose-pink flowers are backed by 2 unequal valves that have withered by anthesis, the longer remnant being inordinately long, 2–2½ times the diameter of the umbel (2.5 cm/1 in). The bell-shaped flowers display long, exserted, pink filaments topped with prominent yellow anthers. Flowering July. HZ 3

The plants described grew from seed received in 1982 as *A. olympicum*, flowering outdoors in 1984 and resembling illustrations of *A. stamineum* in Rix & Phillips, *Bulb Book*. Similar plants but with slightly more erect stems were grown by Jerry Flintoff in Seattle as AC&W 1956, labelled *A. kurtzianum*.

While these species are very similar, *A. kurtzianum* exhibits yellow anthers and cylindrical leaves while *A. olympicum* has violet anthers and flattened leaves.

Adding to the confusion, seedlings of JCA 361 (collected by J.C. Archibald) have proved equally hardy on a raised gritty bed in Cumbria alongside the plants described above. These have paler perianth segments, pale green leaves and are much quicker to produce bulblets but otherwise appear very similar.

Growing *Allium* from seed is a lottery; some unwelcome bulbiferous forms arrive unheralded. The excellent plants that circulate under the collectors' numbers listed as *A. olympicum* restore one's delight in the unexpected. Why such first-class rock garden plants are so little known is quite surprising. The late V. Horton, highly regarded as a connoisseur of bulbs, devoted 25 lines of praise in the *Bulletin of the AGS*, Vol. 45, p. 343, as long ago as 1977.

SOURCES *Fl Turkey*, Section Codonoprasum: *A. kurtzianum*, 54; *A. olympicum*, 60; *A. stamineum*, 64.

A. oreophilum
(= mountain loving)
Meyer   1831

Five–10 cm stems (2–4 ins) grow from

A. oreophilum (Author)

medium-sized, globular bulbs, their tunics grey and papery. The 2 linear leaves, longer than the flowering stalk, flop on the ground but have withered by flowering time. Wide purple bells with dark striping are held, on longish pedicels, in an open umbel, 4–6 cm in diameter (1½–2⅓ ins), spherical or shuttlecock in shape. The stamens are included, the style 3-lobed. In the cultivar 'Zwanenburg' the flowers are carmine red.

Despite inhabiting screes and rocky places around 3,500 m (10,000 ft) in the Caucasus, Pakistan, Afghanistan, Iran and Turkey, blooming in July, *A. oreophilum* is an easy plant for the garden provided sun and good drainage are available. Easily obtainable from popular bulb merchants, usually under the name *A. ostrowskianum*, the bright carmine flowers are pretty, cheerful and quite definitely not magenta (Pl. p. 54). HZ 4

The ordinary form of *A. oreophilum* is less often seen, though quite an elegant flower of more sombre hue.

SYNONYMS *A. platystemon*, *A. ostrowskianum*.

SOURCES *Fl Iranica*, 19; *Fl USSR*, 180; *Eur Garden Flora*, 37. Illus: Rix & Phillips, *Bulb Book*; Hay & Synge, *Dictionary Garden Plants*, 666.

A. oreoprasum
(= mountain leek)
Shrenk   1842

The bell-shaped, rose flowers, in the few- or many-flowered shuttlecock or hemispherical umbel, have a dull purple vein. They sit on equal pedicels up to three times as long as the tepals with their turned back tips. The ribbed stem reaches 20–40 cm (8–16 ins), slightly longer than the 3–5, broad (1–4 mm), channelled, basal leaves. Brown-netted, cylindrical bulbs cluster on a horizontal rhizome. The stamens are shorter than the perianth segments. Pale, flesh-coloured flowers have also been described.

Plants are found on stony slopes and in rocky places throughout Central Asia, Pakistan to West Nepal, 2,700–5,000 m (8,000–15,000 ft), flowering June–July.

A. oreoprasum is rarely seen in cultivation.

SOURCES *Fl USSR*, 35; Hooker, *Fl of British India*, 23; *Fl Rei Pop Sin*, 24. Illus: Polunin, *Flowers of the Himalaya*.

A. oschaninii
Fedtschenko 1906

Wendelbo encountered this onion growing in northern Afghanistan, rooted into steep, rocky crevices in company with *A. giganteum*. Perhaps this sounds an unlikely site to find what many consider the forerunner of our domestic onion.

The bulbs are 2.5–4 cm thick (1–2 ins), certainly smaller than present-day greengrocer stock, with reddish-brown coats, 1–3 at a time attached to a rhizome. Hollow, 45–100 cm (18–40 ins) stems, inflated below their middles, carry spherical, many-flowered umbels. Star-shaped, white segments with a green nerve, 4–5 mm (⅕ in), on 12–20 mm (½–¾ ins) pedicels are backed by a single, lobed spathe. The stamens are a quarter as long again as the tepals. From the base 4 or 5 hollow, upright, glaucous leaves reach one-third the way up the stem.

A native of Central Asia, it is unlikely that many gardeners would wish to grow *A. oschaninii*, save as a curiosity. In the vegetable plot, forms of *A. cepa* will provide a better harvest; furthermore, *A. oschaninii* might be slightly tender for a perennial crop.

SYNONYM *A. cepa* var. *sylvestre*.

SOURCE *Fl USSR*, 92.

A. pallens (= pale)
Linnaeus 1762

A. pallens can be compared to *A. paniculatum*, having similarly a long spathe splitting into 2 unequal valves, the longer 2.5–10 cm (1–4 ins). Unlike *A. paniculatum*, the leaves are thready, 30 cm (12 ins), while the umbel is smaller, 1.5–3.5 cm (⅗–1¼ ins) wide. The flower stalks too are shorter, 5–15 mm (¼–¾ ins), almost equal in length (a marked contrast to *A. paniculatum*), lengthening to 25 mm (1½ ins) in fruit. The flowers are narrow bells, the yellow anthers just peeping beyond the tepal tips, the latter pink or white in colour.

A native of southern Europe, flowering in May to July, *A. pallens* is not as hardy as *A. paniculatum*, needs a summer rest and will require a bulb frame or pot culture in most gardens. V. Horton in *AGS Bulletin*, Vol. 45, p. 344 found it hardy outdoors. In the wild it flourishes on dry grasslands, slopes and rough ground up to 1,800 m (5,400 ft). HZ 4

The following subspecies are described in *Flora Europaea*.

A. p. subsp. *pallens* (syns. *A. paniculatum* var. *pallens*, *A. coppoleri*, *A. amblyanthum*)
Has a compact head of white flowers.

A. p. subsp. *tenuiflorum* (syns. *A. tenuiflorum*, *A. paniculatum* var. *tenuiflorum*)
A rather looser umbel, of usually pink flowers.

A. p. subsp. *siciliense*
Carries pink flowers in a lax head.

SYNONYM *A. serbicum*.

SOURCES *Fl Eur*, 57; *Fl Turkey* 41; *Eur Garden Fl*, 53; *Hortus III*. Illus: Pastor & Valdés, *Revision del Genero Allium en la Peninsula Iberica e Islas Baleares*, 91, 1983; *Bot Mag*, 1420.

A. *paniculatum* (Author)

A. *paniculatum*
Linnaeus 1759

A. *paniculatum* follows the pattern of A. *flavum* and A. *carinatum* subsp. *pulchellum*. A long spathe breaking into 2 unequal valves becomes a trade mark. While this eases the difficulty of species' identification, A. *paniculatum* can appear in various heights and colours. Placing individual plants in correct subspecies may not be as simple.

Membraneous-coated bulbs, 1–2.5 cm in diameter (2/5–1 ins), give rise to 30–70 cm stems (12–30 ins). Three–5 leaves sheathe the lower half, channelled at the upper ends, prominently ribbed on the lower. The spathe, long and beaked, splits into 2 valves, markedly unequal in length, the longer 5–15 cm (2–6 ins). Standing proudly erect as the buds form, when they break free the split valves begin to droop.

The umbel is hemispherical, 3–7 cm (1 1/4–2 3/4 ins), usually packed with flowers on long, unequal stalks, 1–7 cm (2/5–2 3/4 ins), the outer more mature ones drooping, the inner standing erect. The flowers are slightly flared bells with the yellow-anthered stamens usually just level with the tepal edges; bulbils are absent. Tepal colour ranges through apricot, pink, white, yellowish or greeny-brown, often marked on the outside with a darker stripe. Some forms are attractive, subtle shades of old rose, others are rather dingy; the gardener will find great variation in seed-grown stock.

Found throughout Europe through to Central Asia on steppes, hillsides and sandy localities, several forms of A. *paniculatum* have grown for years in the front of a border in Lancashire, without the benefit of much sunshine or drainage. July–September. HZ 1

My plants exhibit a quiet range of subtle shades that I thoroughly enjoy, the individual flowers repaying close scrutiny. Reports of perfume vary, some forms may be fragrant. Others remind a certain grower of a monkey house, another catalogue evokes fried kidneys. Thankfully my plants smell of nothing, not even onions.

Flora Europaea commenting on the very variable forms lists the following provisional classification.

(1) A. *p.* subsp. *villosulum* (syns. A. *paniculatum* var. *villosulum*, A. *rhodopeum*) has a hairy leaf sheath, brown-pink or purplish flowers with darker striping. Greece, Bulgaria.
(2) A. *p.* subsp. *paniculatum* usually has a smooth leaf sheath, lilac-pink or white flowers and grows in most of the range for the species.
(3) A. *p.* subsp. *fuscum* (syn. A. *fuscum*) also has a smooth leaf sheath, greeny-brown or dingy white flowers, sometimes with a pink overlay. Romania, Greece.
(4) While the above three have the spathe-valves at least twice as long as the umbel, A. *p.* subsp. *euboicum* (syn. A. *euboicum*), has valves no more than twice as long. The flowers are yellowish or brownish often streaked with red. Eastern Greece.

A slightly different description of material is found in the *Flora Turkey*.

SYNONYMS A. *longispathum*, A. *praecissum*, A. *paniculatum* var. *legitimum*.

SOURCES *Fl Eur*, 56; *Fl Turkey*, 38; *Fl USSR*, 105; *Eur Garden Fl*, 52. Illus: Rix & Phillips, *Bulb Book*; Grey-Wilson & Mathew, *Bulbs*.

A. paradoxum
Sirak-e zangouleh, Persian

G. Don   1827

A. paradoxum itself need detain no one further, being of little interest other than that of curiosity. The umbel may sport one flower or none having instead a head of bulbils. Originating in the Caucasus and Iran, it has become locally naturalised in Europe and in some areas of Britain.

SYNONYM   *Scilla paradoxus.*

SOURCES   *Fl Eur*, 37; *Fl USSR*, 178.

A. paradoxum var. normale
Stearn   1987

In the last few years bulbs have been circulating, introduced in 1966 by Rear Admiral Paul Furse from northern Iraq, which are excellent garden plants. Not until 1987, however, was *A. paradoxum* var. *normale* described in the *Kew Magazine* by Professor Stearn.

Ovoid bulbs with membraneous coats, 1.5 cm in diameter (3/5 ins) send up a single, curved, gleaming, bright green leaf in early spring to 33 cm × 2.5 cm (13 ins × 1 ins)). Like *AA. triquetrum, pendulinum* and *ursinum* the leaves grow with their inner surfaces curving inwards, so that the keeled surface of *A. paradoxum*, which one would expect to find on the undersurface of the leaf, has curled over to appear uppermost. From the top of a 3-angled stem, 15–30 cm (6–12 ins), hang 10–15 nodding, white, bell-shaped flowers with broad tepals, to 11 mm long × 6.5 mm wide. The flower stalks are half as long again as the perianth segments, 2–4.5 cm (3/4–1 3/4 ins). The stamens tucked inside the flower, and the 3-lobed stigma are white. (Pl. p. 7.)

A. paradoxum var. *normale* is as welcome in spring as Snowdrops and Lily-of-the-Valley, whose colouring it echoes. The tepals are a crisp gleaming white, large enough to have impact at a border front. The leaves remain fresh and succulent until after seeding. Plants enjoy the heavy, wet clay of our Lancashire garden and flourish better outside than in pots. Already odd seedlings have appeared in crevices of flagged paths, welcome as potential gifts. The seeds have an oil-bearing appendage, attractive to ants, which aids dissemination. HZ 4

Plants shown by Mrs Felicity Baxter, who had received bulbs from Admiral Furse, gained an Award of Merit, noted in the *AGS Bulletin*, Vol. 49 and were illustrated in Vol. 48, p. 314.

SOURCES   *Fl Eur*, 37; *Fl USSR*, 178; *Eur Garden Fl*, 42. Illus: Rix & Phillips, *Bulb Book*; *Kew Magazine*, Nov 1987; *Fl Iranica*.

A. pendulinum
Tenore   1810

This was classified by Regel as a variety of *A. triquetrum*. However, the umbel is not 1-sided in *A. pendulinum* as it is in the former.

From the almost spherical bulbs the leaves clothe the stem with a very short sheath, appearing almost basal and markedly keeled. The 3-sided stem, 6–25 cm, (5–10 ins) supports an umbel with 3–5 flowers, the flower stalks 4 cm (1½ in), drooping on maturity. Attractive, star-shaped, white flowers carry a green stripe on each tepal, 3–5 mm × 1–1.5 mm (¼ × 1/20 ins). According to *Fl Europaea*, however, there appear to be forms with tepals almost half as large again.

Found in shady, damp locations and woods in central Mediterranean areas, it should flourish in the same environment as *A. triquetrum*, or as a pretty and effective pot plant. (Pl. p. 61.) HZ 4

A. *peninsulare* (Author)

SOURCES   *Fl Eur*, 36; *Eur Garden Fl*, 41. Illus: Reichenbach, *Icones Florae Germanica*, 10: t.503.

A *peninsulare* Peninsular Onion
Lemmon   1888

Another attractive Californian onion, with deep rose-purple flowers that hold their colour on the tepal ends for at least five years when dried. A larger plant than A. *crispum*, which some authorities have considered to be a varietal form.

Horizontal fissures on the grey tunics of the oval bulbs crack when dry. The flat, 2–4 leaves equalling the length of the scape, 20–30 cm (8–12 ins), have withered by anthesis. An open umbel carries 18–20 flowers on long, spreading stalks, 2–3 cm (¾–1½ ins). While the outer tepals are acuminate, the narrower inner ones are more erect, with fine crimping along the margins.

A. *peninsulare* can be found blooming from March to June on dry wooded slopes below 1,000 m (3,000 ft) in California. A pretty onion well worth growing; in cultivation a bulb frame or alpine house is required to provide summer rest. HZ 4

SYNONYM   A. *montigenum*.

SOURCES   Munz, *California Flora*. Illus: Rickett, *Wildflowers U S*, Vol. 4; Abrams, *Illus Fl Pac States*; *AGS Bulletin*, Vol. 45, p. 340.

A. *polyastrum* (= many-starred)
Diels   1912

One of George Forrest's introductions, discovered in the Lichiang Range in western China, A. *polyastrum* grows above 3,000 m (12,000 ft) in damp, grassy areas. This is a good garden plant for the open border, flowering August–early September.

The stems are strongly angled, often producing 4 marked, vertical ribs, some years contorting into strange arabesques, in others held stiffly upright. The wide, hemispherical heads, 6–8 cm (2½–3 ins), are filled with star-shaped, magenta-purple flowers whose narrow tepals reflex with maturity and pleasantly contrast with the prominent green ovaries. The whole umbel has a firm texture and cuts well for the house. Without doubt, this is a good plant for a perennial border; we have found self-seeding to be no problem in north-west England. (Pl. p. 162.)

The numerous leaves bunch at the base of the stem, sheathing the lower end. The fibrous roots divide easily, showing little evidence of bulb formation. HZ 2

Although this *Allium* is grown in British gardens, references are scarce. It appears in *Flora Reipublicae Popularis Sinicae*, 1980, as a synonym of A. *wallichii*. As the text is in Chinese, there are certain difficulties evaluating the similarity.

In '*Allium* and *Milula* in Central & Eastern Himalaya', p. 182, Professor Stearn allies the two species discussed with A. *lancifolium* and A. *platyphyllum*. Unfortunately, the use of the name *platyphyllum* had already been made in connection with an American onion, A *tolmiei*; an indication of how helpful an updated monograph on *Allium* would be.

At present this is a plant available to

gardeners which is most difficult to track down in the literature, a situation probably unlikely to trouble many of those who are just interested in growing a good garden plant.

SOURCE *RHS Journal*, April 1970, Part 4.

A. porrum (*pirasa* = leek, Turkish)
Leek
Linnaeus 1753

The Leek derives from *A. ampeloprasum*, but over the centuries of cultivation has developed into a sufficiently distinct plant. While leeks are easily identified on a plate by the diner, when allowed to 'bolt' in the vegetable garden, flowers appear which may confuse the grower.

The bulb is scarcely developed appearing merely as a slight bulge above the root plate. The broad, flat leaves sheathe the stout and solid stem which unchecked soars up to 1 m (36 ins), to be capped by a spherical umbel to 9 cm (3½ ins) diameter, with a single-valved spathe which withers fairly early. Hundreds of pinkish or whitish cup-shaped flowers with marginally exserted stamens on slightly unequal pedicels, pack the flower head and can be picked for large flower arrangements. For cultivation, see p. 22. June–July. HZ 2

SOURCES *Fl Eur*, 75; *Fl Turkey*, 77; *Fl USSR, 176.*

A protensum
(tensum = stretched out)
Wendelbo 1968

Plants of *A. schubertii* growing in Central Asia were described by Vvedensky in *Flora USSR* in 1935, the general distribution being the eastern Mediterranean, in particular Palestine.

Professor Per Wendelbo studied plants from both regions and having decided that the material from Central Asia belonged to a separate taxon, he included the latter in *Flora Iranica* as *A. protensum*.

Papery, blackish bulbs, 2–3 cm (¾–1¼ ins) in diameter, give rise to a 10–25 cm (⅖–1 ins) stem. Widish leaves 1–5 cm (⅖–2 ins), longer than the scape, carry tiny teeth along the margins. The flower head, backed by a 3-valved bract, is spherical, loosely packed with bell-shaped flowers on stalks that vary greatly in length. Those bearing fertile flowers may be 6–16 cm (2 ⅓–6⅓ ins), while the sterile blooms have pedicels twice as long, the whole incredible umbel being capable of exceeding 30 cm (12 ins). The tepals are pale brown with darker mid-veins, appearing dull white with purplish nerves when dried.

Flowering in May–June, Afghanistan, Iran, Central Asia at 300–3,000 m (900–9,000 ft). Plants are cultivated at the Royal Botanic Gardens, Kew.

This species is still very hard to obtain, propagation being primarily by seed, with a long wait—in the region of five years—before flowering size is attained. Mere size will increase desirability, with the challenge of cultivation also appealing to the acquisitive nature of the confirmed *Allium* buff.

In *The Bulb Book*, *A. schubertii* is described as frost tender; *A. protensum* as possibly quite hardy in dry conditions. For the average grower, cultivation may well demand a bulb frame or alpine house, with a summer rest. HZ 4

SYNONYM *A. schubertii* sensu Vvedensky, *Fl USSR.*

SOURCES *Eur Garden Fl* 79. Illus: *Fl Iranica*, 129; Rix & Phillips, *Bulb Book.*

A. pseudoflavum
Vvedensky 1934

A. pseudoflavum mainly differs from *A. flavum* in having narrow grass-like leaves, only 0.5 mm wide, with the lower sheaths and leaf blade minutely hairy, compared with the 2 mm, slightly grooved, terete, hairless leaves of the latter. The bulbs are lighter in colour with outer, greyish coats, the inner being yellowish, leathery and with distinct veining. *A. flavum* has bulbs with blackish, membraneous tunics. Both have predominantly yellow flowers on long unequal pedicels, no bulbils and long spathes.

Plants are found on dry slopes and roadsides in northern Iran, Caucasus around 300–2,000 m (900–6,000 ft), flowering in June to August. HZ 4

SYNONYMS *A. filifolium, A. freynii, A. freynianum.*

SOURCES *Fl Turkey,* 56. Illus: *Fl USSR,* 101.

A. pskemense
Fedtschenko 1905

These Onion-like plants appear in seed lists, having reddish-brown bulbs on a rhizome, tall, hollow stems, 40–80 cm (16–32 ins), inflated below the middle, with basal leaf sheaths, the 3 cylindrical leaves being hollow and half the length of the stem. The packed, spherical umbel, with a spathe the same length as itself, carries white, stellate flowers, with slightly exserted stamens, on long, equal pedicels. Flowers in August, in stony places, Central Asia. HZ 3

SOURCE *Fl USSR,* 90.

A. pulchellum

This is an inaccurate name for *A. carinatum* subsp. *pulchellum*. Gardeners will have no doubt which title trips off the tongue more readily and it is quite possible that *A. pulchellum* will continue to be used, in commerce at least, for some time to come. For this reason alphabetical order has been bent, conforming with popular usage.

A. carinatum subsp. *pulchellum* 'Album' in Hartsop, Cumbria (Author)

A. carinatum subsp. **pulchellum**
Bonnier & Layens 1898

The 30–60 cm (12–24 ins) stem, rising from brownish, membraneous bulbs, carries 2–4 sheathing, linear leaves with prominent veins. The spathe is greatly elongated into a tapering beak, erect while the flower buds form. These break clear, splitting the spathe into 2 valves, unequal in length, the longer up to 12 cm (4¾ ins). These gradually droop, still being present at fruiting.

The purple stalks are long, 10–25 mm, (⅖–1 ins) and unequal; the outer ones carrying the more mature flowers curve downwards, turning the umbel into a falling fountain of purple-rose, bell-shaped elegance. Beyond the tepal edges the long, purple stamens are tipped with yellow pollen. A soft bloom overlays all the flower head. As the flowers are fertilised they once more stand erectly.

Several forms of *A. carinatum* subsp. *pulchellum* exist, some with glaucous leaves. Dwarf plants from the mountains make suitable rock garden inhabitants and a very lovely clear, clean white form comes true from seed. I grow a most satisfying clump of intermingled rose-purple and white flowers.

One selection of plants flower in July and August, set plenty of seed and keep their leaves green for a longer period than another group I have had for about 15 years. With the other not a single seedling has appeared, the plants are taller, darker in tone, flower in August and September and are slow to produce offsets. Professor Stearn describes a form that is very similar with yellow anthers that he calls, *A. carinatum* subsp. *pulchellum* van Tubergen's variety. My plants have yellow anthers on the young flowers, later darkening with age.

Cultivation in the open garden is easy, plants only asking not to be overlaid by dominant neighbours or planted in over-wet ground. The leaves are withering by flowering time, this can be concealed by judicious placing. Remove seed heads on the exuberant forms, old flowering stems come away neatly and as autumn approaches new leaves will begin to appear. *A. carinatum* subsp. *pulchellum* in all its forms is one of the most pleasing alliums to grow.

Found in southern Europe and Asia Minor, heaths, rocky ground, June–September. HZ 3

SYNONYMS *A. pulchellum* Don, nom illegit, *A. flavum* var. *pulchellum*.

SOURCES *Fl Eur*, 71: Grey, *Hardy Bulbs*. Illus: Hay & Synge, *Dictionary of Garden Plants*, pl. 667; Blanchard, *JRHS*, Vol. 32, p. 170.

A. 'Purple Sensation'

Featured at Chelsea Flower Show in 1988, the 10 cm (4 ins) wide heads of rich, reddish-violet on 90 cm (36 ins) stems resemble *A. stipitatum* rather than *A. aflatunense*. For the present its parentage remains confused.

Shown by the Director, Royal Botanic Gardens, Kew, as a hardy bedding plant also suitable for cutting, the colour may vary a little, while bulbs may prove hardy in well-drained areas in sunny gardens, but new introductions take time for full evaluation. HZ 2–3

SOURCE *The Garden*, pp. 216–17, Dec 1988.

A. pyrenaicum
Costa & Vayreda 1877

Ovoid bulbs with membraneous coats produce 55–100 cm (22–40 ins) stems. Five–6 flat, keeled, linear leaves sheathe the stem for 1/2–1/4 its length, the leaf margins having tiny teeth. The single spathe is long and beaked, falling off before the flowers wither. Umbels, 4–7 cm (1 1/2–2 3/4 ins) in diameter, are packed with dullish white, star-shaped flowers on equal 2.5 cm (1 in) stalks. Tepals sport a green mid-vein and are also margined with tiny teeth; the stamens being about the same length with brown anthers.

Not surprisingly the plant hails from the Pyrenees, among rocks around 900–1,200 m (2,500–3,500 ft). Rarely if ever occurring in gardens, cultivation would probably cause little difficulty. HZ 2

A. pyrenaicum is included, not on its own merits, but because plants appear in commerce incorrectly under this name. The false *A. pyrenaicum* is much smaller with pale lilac flowers and in most cases is probably *A. angulosum* (qv).

SOURCES *Fl Eur*, 86; *Hortus III*; *RHS Dictionary*.

A. ramosum
Linnaeus 1753

A. ramosum and *A. tuberosum* have many similarities which may create difficulties in identification. Both have cylindrical bulbs with netted coats, clustered on short rhizomes, and 4–9 leaves about 35 cm (14 ins) long. While those of *A. ramosum* are semicircular, hollow and sheathing the stem for one-third its length, on *A. tuberosum* they are solid, keeled and sheathing for only one-eighth. The umbel of *A. ramosum* is compressed into a funnel, with few to many, bell-shaped, white flowers with a dark red stripe. On the contrary the umbel of *A. tuberosum* is usually hemispherical (sometimes compressed) with numerous starry, white flowers underlined with faint green or brown. Both have umbels measuring 3–5 cm (1¼–2 ins) and stems of 25–50 cm (10–20 ins). The tepals of *A. ramosum* are 8–10 mm (⅖ ins), twice the length of the stamens, the anthers purple-brown with yellow pollen. Plants are sweetly scented.

A native of Central Asia, plants are easily grown in the open garden. *A. ramosum* blooms in June–July, earlier than *A. tuberosum* which flowers in late July–September (Fig. p. 57). HZ 3

No little degree of confusion surrounds the synonomy of both plants. It appears that the ageing Linnaeus confused the living plant he named *A. odorum* with the herbarium specimen he had named earlier as *A. ramosum*. Later Regel included two separate species which are now recognised as *A. ramosum* and *A. tuberosum* under the name *A. odorum* L. As the name *A. odorum* was also given by Thunberg in 1784 to *A. thunbergii* Don, one can only conclude that a sweet-smelling onion goes to a botanist's head! (Stearn, *Nomenclature & Synonymy of* Allium odorum *& A.* tuberosum. Herbertia, 1944.)

A. tataricum and *A. tartaricum* have also been noted in connection with *A. inderiense* and *A. barszczewskii*.

SYNONYMS *A. odorum*, L, *A. tataricum* Linnaeus filius, *A. potaninii*, *A. weichanicum*.

SOURCES *Eur Garden Fl*, 17; *Fl Rei Pop Sin*, 26; Grey, *Hardy Bulbs*. Illus: *Bot Mag*, 1142 (1808) (*A. tartaricum*); Rix & Phillips, *Bulb Book*.

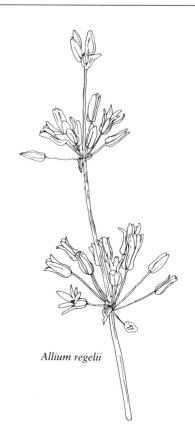

Allium regelii

A. regelii
Trautvetter 1884

Illustrations of this unusual onion abound but plants are exceedingly rare in cultivation. The reason for the interest shown in the species is the layered series of umbels on the flower stem, with the plant resembling a Candelabra *Primula* rather than an *Allium*.

Found in Iran, Afghanistan, through Central Asia, in rocky sites, from 700–2,450 m (2,000–7,200 ft), plants have purplish flowers, rarely nearly white, and some may have only one tier to the umbel.

Despite being described as long ago as 1884 and re-introduced by Admiral Furse in more recent times, *A. regelii* is not easily available. Brian Mathew reports that it sets seed, requires a summer rest and should not prove too difficult to grow. For the moment it remains a species to look forward to.

SYNONYMS *A. yatei*, *A. cupuliferum* var. *regelii*.

SOURCES *Fl USSR*, 244; Mathew, *Dwarf Bulbs*. Illus: *Fl Iranica*, 139; Rix & Phillips, *Bulb Book*.

A. robinsonii
Henderson 1930

A most interesting account of this little onion has been written by Mark McDonough, highlighting the ease with which plants having a highly specialised location may become extinct through changes wrought on the landscape by human interference.

In this instance, a small *Allium* restricted to low, sand banks on the Columbia River, from Vantage, Washington to John Day River, Oregon is threatened by water level changes. Indeed the entire flora of the Columbia River Gorge may be at risk if the water level is altered in the name of progress.

Around Vantage the deserts of east Washington run up to the verges of the enormously wide river. Indian pictographs can be seen along the cliffs, petrified tree remains and a wealth of desert plants are there for the visitor. Finding *A. robinsonii* required a prolonged search, then in May 1984 came success.

Cultivation in pots or dry, raised, sunny beds has provided flourishing plants, allowing seed to be circulated. *A. robinsonii* has a chance now to escape extinction.

Tiny, winged scapes, 2–8 cm (¾–3 ins), 2 sickle-shaped leaves, pale pink flowers with blackish-red stamens suggest an attractive allium for a trough or alpine house. April–May.

The fate of *A. robinsonii* suggests how an individual can make a personal contribution to the preservation of threatened species.

SOURCES Illus: McDonough, *RHS Lily Group Year Book*, 1986–7; Hitchcock et al., *Vasc Plants Pac N W*.

A. rosenbachianum
Regel 1884

Dark, papery, nearly spherical bulbs, 1.5–2.5 cm (¾–1 ins), produce a 30–100 cm (12–40 ins), noticeably ribbed stem with 2–4 much shorter, basal, bright green leaves. Below the loose, many-flowered, spherical umbel lies a shorter spathe. Dark violet, star-shaped flowers with even darker mid-veins top unequal pedicels, 2–6 cm (¾–2½ ins) long. Stamens 1½ times the perianth length project beyond the tepals, which twist and reflex after maturing. The rather nodular ovary sits on a short stalk.

Afghanistan, Central to south-west Asia on stony slopes, 2,000–3,500 m (5,000–

Allium robinsonii

10,500 ft) and, according to *Flora USSR*, 212, growing in the shade of rocks and trees. *A. rosenbachianum* flowers in May. HZ 4

First collected near Bokhara, according to Regel this was a superb species. In gardens a warm, well-drained border will be required; the species is reputed to grow freely from seed. Bulbs are easily obtained from specialist nurseries, though *A. stipitatum* has sometimes been supplied instead.

A white form is available commercially.

SYNONYM   *A. angustitepalum.*

SOURCES   Grey, *Hardy Bulbs*; *Eur Garden Fl*, 71; Mathew, *Dwarf Bulbs*. Illus: *Fl Iranica*, 118.

A. roseum
Linnaeus   1753

Numerous bulblets surround the round bulb with its hard brittle coat, pitted with minute perforations. Two–4 linear leaves sheathe the lower end of the solid, 10–65 cm (4–26 ins) scape. A persistent spathe's deeply lobed, single valve backs the hemispherical umbel up to 7 cm (2¾ ins) in diameter. Variable proportions of bulbils and flowers can be found in the flower head: pale pink or white, bell- or cup-shaped flowers with 7–12 mm (¼–½ ins) long segments enclose shorter yellow-anthered stamens.

A. r. var. *roseum*
Many-flowered, fertile, without bulbils.

A. r. var. *bulbiferum*
Sterile, with mainly bulbils.

A. roseum is found throughout southern Europe, Turkey and North Africa in dry, open spaces and on cultivated ground. Due to the ease with which plants propagate via bulbils and bulblets, numerous clones have received names: *AA. ambiguum, carneum, amoenum, incarnatum, majale,* as well as *A. roseum* subsp. *bulbiferum* and vars. *bulbiferum, bulbilliferum* and *carneum. A. confertum* is a dwarf variant from Corsica and Sardinia (syns. *A. roseum* vars. *insulare* and *humile*). Flowering April to June. HZ 4

Invasive, bulbiliferous forms are unlikely to be very welcome in the garden; care should be taken to obtain plants with good flowers. *A. roseum* is then a handsome species with broad overlapping tepals, usually of an attractive shade. Cultivation is easy in warm, well-drained areas.

SOURCES   *Fl Eur*, 23; *Eur Garden Fl*, 28; *Fl Turkey*, 15. Illus: *Bot Mag*, 978 (1806), *A. roseum bulbiferum*; Rix & Phillips, *Bulb Book*; Polunin, *Concise Flowers of Europe, Fl Greece & Balkans.*

A. rubens
Schrader   1809

Slightly ribbed, slender stems, 10–25 cm (4–10 ins), rise from narrow, conical bulbs crowded on to a rhizome surrounded by 5–6 grooved, semi-solid, thready, basal leaves. A single, short spathe backs the globular or hemispherical umbel with its few lax flowers on even stalks. Purplish, broadly bell-shaped flowers have segments, yellow-anthered stamens and style roughly all 4–5 mm (⅕ ins) long.

Plants are found flowering from June to August on steppes and rocky slopes from the southern Urals through north & central Asia to Siberia. Cultivation should create little difficulty in the open garden. HZ 2

SYNONYM   *A. stellerianum.*

SOURCES   *Fl Eur*, 2; *Fl USSR*, 42; *Hortus III*; *RHS Dictionary.*

A. rupestre
Steven 1812

From the group with a long-beaked spathe breaking into 2 unequal valves, *A. rupestre* is hardly spectacular. Yet being almost the last *Allium* to flower before the winds and rains of autumn drive the Cumbrian gardener indoors, *A. rupestre* is cherished for surviving on the rock garden these last six years.

Resembling *A. pallens* and *A. paniculatum*, the spathe of *A. rupestre* measures up to 5 cm (2 ins). While the rather stocky stem is 25–40 cm (10–16 ins), the 2–3 threadlike leaves forming a sheath around the lower half are only 12 cm (4¾ ins). The umbel, 2–3 cm wide (¾–1¼ ins), hemispherical or shuttlecock in shape, carries 20 flowers or less. Narrow, bell-shaped flowers of white or dusty pink, the mid-vein purple or pink, are held on almost equal stalks, 5–10 mm (¼–½ ins). While the style extrudes, the stamens are about tepal length, tipped with violet anthers.

No bulbils are formed; plants grow easily from seed, flowering in their second year. Cultivation should hold few problems in the open. Found on stony slopes in the Crimea and Caucasus, flowering August to October (September–October in Cumbria). HZ 1–2

SYNONYMS *A. paniculatum* var. *macilentum*, *A. p.* var. *pallens*, *A. p.* var. *rupestre*, *A. charaulicum*, *A. subquinqueflorum*, *A. tristissimum* and ? *A. triste*.

SOURCES *Fl USSR*, 109; *Fl Eur*, 59; *Fl Turkey*, 43; Grey-Wilson & Mathew, *Bulbs*. Illus: Reichb, *Icones Florae*, 5.

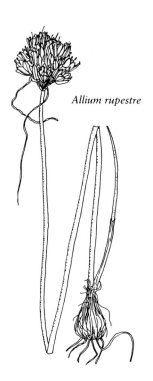

Allium rupestre

A. sativum
(sativus = cultivated)
Garlic
Linnaeus 1753

Such a well-known bulb hardly requires description. The depressed ovoid in its papery, white skin contains 5–15 bulblets more familiar in the kitchen as cloves. The flowering stem, 25–100 cm (9–40 ins), with its 6–12 leaves sheathing the stem, is more rarely seen. In fact the plant has quite an imposing architectural appearance, growing in warm sunny corners. Like a perky nightcap a long, beaked spathe tops the umbel, which carries firm, rather yellowish bulbils and very few flowers. These are greenish-white, purple or pink, the stamens included or just equalling the tepals.

In sunny climates, garlic can decorate flower gardens not only the vegetable patch. A clump in Victoria, Vancouver Island, formed quite a striking backing to a mixed border. HZ 4

Unknown in the wild, *A. sativum* is probably represented by *A. longicuspis*, distinguishable from garlic by stamens longer than the tepals, if the pursed flowers open sufficiently to show them.

The plant probably most correctly known as *A. s.* var. *ophioscorodon*, but appearing in Clusius' *Rariorum Plantarum Historia* (1601) as *A. controversum*, has a quite extraordinary stem, which before flowering loops the loop. A Plate of the coiled stem appears in Clusius, *Rariorum Plantarum Historia* (1601).

SOURCES *Fl Eur*, 75; *Fl Turkey*, 89; *Fl USSR*, 163. Illus: Reichenbach, *Icones Florae Germanicae*, 10: 488 (1848).

A. saxatile
Bieberstein 1798

While resembling *A. senescens* (qv), *A. saxatile* differs in its narrow bulbs, threadlike leaves sheathing the lower third of the stem and a spathe which splits into 2 valves, one longer than the other, to 3 cm (1¼ ins), and also exceeding the umbel. Deep pink to white flowers, with long protruding stamens, present in a 2–4 cm (¾–1½ ins), almost spherical flower head.

Deeper-coloured plants with pink flowers and purple anthers grow in the eastern Alps towards Central Russia, while in Crete, northern Italy and into Yugoslavia forms appear with yellow-white flowers and yellow anthers.

A. saxatile (Author)

Growing on rocky places and dry plains, flowering in July–August, a dry, well-drained area of the rock garden will be adequate. HZ 3

Flora USSR treats *A. globosum* as a separate species.

SYNONYMS *A. globosum, A. marschallianum, A. savranicum, A. caucasicum, A. stevenii* var. *ruprechtii*.

SOURCES *Fl Eur*, 4; *Fl USSR*, 71.

A. scabriscapum
(scabrid = rough to the touch)
Sirak -e chap piaz, Persian
Boissier & Kotschy 1853

The only yellow *Allium* species in Iran, plants are quite common in the Zagros and Alborz ranges, recorded in Central Asia, the Caucasus, Turkey and Iraq, on gravelly slopes around 1,850–1,900 m (5,500–6,000 ft). Infrequently grown, the attractive plants at the Royal Botanic Gardens, Kew should stimulate demand.

Narrow, conical, dark, fibrous-netted bulbs, clustered on a short rhizome, produce a 15–50 cm, finely ribbed stem (6–20 ins). Four–6 shorter, glaucous, linear, flat leaves can, like the scape, be hairy or bare. Globular or sometimes hemispherical open heads of attractive, clear yellow flowers, backed by a 2–4 lobed spathe, bloom in June/July. Bell-shaped flowers pack the broadish umbel, 3–4.5 cm (1¼–1¾ ins). Long, equal pedicels, 1.5–2.5 cm (⅗–1 ins), dry to almost black in contrast to bright, green-striped tepals bleaching to off-white. While the stamens approximately equal the tepal length, the style is quite prominent.

A. scabriscapum is reputed quite difficult to grow; a challenge, for this is a plant with attractive colour. Frame or alpine house can provide the necessary summer rest. HZ 4

SOURCES *Fl Iranica*, 1; *Eur Garden Fl*,

6; *Fl USSR*, 21. Illus: Wendelbo, *Tulips and Irises of Iran*.

A. schoenoprasum
(schoen = rush; prasum = leek)
Chives
Linnaeus 1753

Numerous slender, membraneous, conical bulbs cluster on short rhizomes. One–2 hollow, cylindrical leaves, to 35 cm (14 ins), sheathe a third of the scape. This too is hollow and ranges from 5–50 cm (2–20 ins), under the variably lobed, short spathe. Densely packed flower heads, 1.5–5 cm (3/5–2 ins) in diameter, may carry as many as 30 flowers on 2–20 mm, slightly unequal stalks. Bell-shaped flowers, of pink, lilac, medium purple or white, bear tepals from 7–15 mm (2/5–3/5 ins) long with included stamens. Many local variants.

Ubiquitous across the northern hemisphere; in Alaska, Eskimos kept plants for long periods in airtight sealskin bags. To survive over half the globe while looking pretty, even uncooked, is a triumph of survival. Cultivation is very simple, almost every garden has a clump of Chives, plants are able to flourish in shade and damp peat beds. May–August. HZ 1

A. s. 'Forescate'
Even in the kitchen garden, good forms should be given preference. This has deep clover pink heads on stems 20–25 cm (8–10 ins). Seed does not come true but some seedlings appear that are a deep purplish-lilac. Clear white forms are of chives, to my knowledge there are no white *A. s.* 'Forescate' (Pl. p. 55).

Removing seeding heads and feeding the clumps can result in three flower crops a year. Cut flowers emit no smell unless bruised.

Numerous *Allium* resemble *A. schoenoprasum*; no one who has ever eaten chopped chives in a sandwich will forget that stems and leaves are hollow, a useful tip when finding a solid-stemmed look-alike. *A. schoenoprasum* turns up in seed lists with a hundred aliases.

References appear in most floras; illustrations in innumerable vegetable and flower guides.

SYNONYMS *A. sibiricum*, *A. buhseanum*, *A. purpurascens*, *A. raddeanum*.

SOURCES Illus: Rix & Phillips, *Bulb Book*; Grey-Wilson & Mathew, *Bulbs*; Synge & Hay, *Dictionary Garden Plants*.

A. schubertii
Zuccarini 1843
A. schubertii was described by Vvedensky in *Flora USSR* 1935, with *A. bucharicum* as a synonym. By 1963 he had decided they were separate taxa, so *A. bucharicum* appeared in *Flora Iranica* as a distinct species.

Further confusion arises. Plants from Palestine and the eastern Mediterranean were thought by Wendelbo to be different from material from Central Asia. The latter was then placed in a separate taxon, namely *A. protensum*. *A. schubertii* has a longer stem and purplish flowers, while *A. protensum* with buff flowers is shorter.

The bulbs may be ovoid or spherical with brown, leathery coats. The 4–8 bright green leaves, broad and wavy, 6 cm (2⅓ ins) wide, with tiny teeth on the margins, are slightly shorter than the hollow stem, 30–60 cm (12–24 ins). Each spathe splits into 2 or 3 valves below the many-flowered umbel with its striking formation. Starry, violet, white or pink flowers unequal, purplish pedicels, between 2 and 20 cm (3/4–8 ins) long; a sparkling firework design of tepals growing rigid with age. Sterile and fertile flowers are found in the umbel, the sterile usually on longer stalks. The mauve filaments with their yellow anthers are not exserted.

Before seeds have formed, the lower stem has withered, becoming brittle. The

whole head snaps off, a skeleton football rolling away in the wind to scatter seed over fresh ground—an Old World Tumbleweed.

Spring-flowering in the eastern Mediterranean and parts of North Africa, *A. schubertii* needs summer dormancy in a bulb frame or an alpine house. Plants are frost tender. HZ 5

SOURCES *Fl Turkey*, 141; *Fl Iranica*, 90; *Eur Garden Fl*, 78. Illus: *Bot Mag*, 7587, 7588; *Fl USSR*, 219, Pl. XVI; Wilde-Duyfjes, *Revision of the genus* Allium L. *(Liliaceae) in Africa* (1976) and in *Belmontia*, 7:1–237 (1977).

A. scorodoprasum
Sand Leek, Rocambole
Linnaeus   1753

The tawny, friable coats of the ovoid bulbs break away revealing not only dark violet undercoats but, frequently, violet-black bulblets. Two–5 short leaves sheathe the stem, 25–90 cm (9–36 ins), for one-third its length. The single valved, 1.5 cm (⅗ ins) long spathe is beaked and non-persistent. Above, the 1.5 cm (¾ ins) wide umbel is packed with bulbils plus a few lilac or deep purple flowers on unequal stalks. Yellow-anthered stamens are included.

Definitely not a plant to allow into the garden, a self-perpetuating thug flowering in early summer, *A. scorodoprasum* ranges through Europe, to Iran and the Caucasus. HZ 3

Both *A. vineale* and *A. carinatum* have bulbils and may be confused with each other (see relevant entries).

SOURCE   Illus (for comparison): Fitter & Fitter, *Wild Flowers of Britain & N Europe*.

A. s. subsp. *scorodoprasum*
Has bulbils and a few flowers, reputedly sterile. However, some of the subspecies of *A. scorodoprasum* are good plants for the garden.

A. s. subsp. *jajlae* (syn. *A. jajlae*)
Has numerous rosy-violet flowers but no bulbils. Comes from mountain slopes in the Crimea and the Caucasus. Pretty and easy to grow in sunny, well-drained borders. HZ 3

SOURCE   Illus: Rix & Phillips, *Bulb Book*.

A. s. subsp. *rotundum* (syn. *A. rotundum*)
Also lacks bulbils. The outer tepals are dark purple; the inner have a central purple stripe and broad whitish edges.

SOURCES   *Hortus III*. Illus: Polunin, *Fl Greece & Balkans*, Pl. 56.

A. s. subsp. *waldsteinii* (syns. *A. waldsteinii*, *A. rotundum* subsp. *waldsteinii*)
Has uniformly dark purple tepals.

SOURCES   *Fl USSR*, 170; *Fl Eur*, 87; *Fl USSR*, 161.

A. scorzonerifolium
De Candolle (DC)   1804

A. scorzonerifolium resembles *A. moly* (qv) having star-shaped, yellow flowers but differing in its narrower leaves.

A. s. var. *scorzonerifolium*
Has bulbils in the head and few flowers.

A. s. var. *xericiense*
Has flowers only.

(Top left:) *A. scorodoprasum* ssp. *jajlae* (Author)
(Bottom left:) *A. scorzonerifolium* (Brian Mathew)

West & central Spain and north-west Portugal, flowering June, July in rocky ground. HZ 4

SOURCES *Fl Eur*, 34; *Eur Garden Fl*, 36; Pastor and Valdes, *Revision del genero Allium*; Illus: Redouté, *Liliacées* 2: t.99 (1804).

A. senescens Mountain Garlic
Linnaeus 1753

Bottle-shaped, narrow bulbs clustered on rhizomes produce 7–45 cm (2¾–14 ins), angular stems, 2-sided at their tops. Numerous slightly shorter, linear, basal leaves have slightly rounded, not keeled, undersurfaces. A persistent 2–3-lobed spathe backs the hemispherical umbel, 2–5 cm in diameter, packed with flowers on 8–20 mm stalks. Each lilac flower is cupshaped, the stamens exserted, the ovary deeply 3-lobed.

Flowering July to September, in dry rocky places, Europe and Central Asia as far as Siberia. HZ 1

There are two broadly geographical subspecies, though the *Flora USSR* describes four.

A. s. subsp. *montanum*
Found from the Ukraine through to northern Portugal; the plant described above.

A. s. subsp. *senescens*
From northern Asia; has broader leaves.

Synonyms appear in seed lists and in commerce. To add to the confusion, intermediate plants occur. One, glaucousleaved, appears as *A. glaucum*. Another, common on rock gardens, with spirally twisted leaves, is the synonymic *A. spirale*.

Gardeners will find variants of *A. senescens* virtually impossible to distinguish.

A. senescens (Author)

Allium senescens subsp. *montanum*

Very similar are *A. angulosum* and *A. suaveolens* but both these have keeled undersurfaces to the leaves. All are easily grown in the border or rock garden but the colour is undistinguished lilac, though the fresh-looking leaves improve the appearance of clumps.

The synonymic form known as *A. spirale* is a delightful small plant for the rock garden, flowering in late July–August when very little still blooms. Unmistakable are the glaucous, spirally twisted leaves and the surface-based rhizome, aping a diminutive Iris. Strongly exserted stamens and tepals of bright pink add to its charm. This plant appears in nurseries labelled *A. glaucum*, or *A. senescens* var. *glaucum*. Looking so different from other forms of *A. senescens*, what a pity a simple name like *A. spirale* is not admissible.

A. senescens var. glaucum in Hartsop, Cumbria
(Author)

A very helpful, simplified list of the comparative features of other similar species, i.e. *AA. lineare, rubens, kermesinum, palentinum, saxatile* and *hymenorhizum*, appears in Grey-Wilson & Mathew, *Bulbs*.

SYNONYMS *A. montanum, A. lusitanicum, A. fallax, A. sensescens* var. *calcareum, A. baicalense, A. spirale, A. glaucum, A. andersonii, A. spurium, A. angulosum* var. *minus*.

SOURCES *Fl USSR*, 44; *Fl Eur*, 3; *Eur Garden Fl*, 1. Illus: *Bot Mag*, 1150; Grey-Wilson & Mathew, *Bulbs*.

A. siculum

Now placed in *Nectaroscordum*, as *N. siculum* subsp. *siculum*, gardeners will take many years to forget the old name.

Above long, fleshy, basal leaves, the greenish-red flowers droop on their tallish stems, 50–125 cm (20–42 ins). After fertilisation the erect seed heads dry to deep cream; greatly prized by flower arrangers.

Found flowering in damp shady woods in southern Europe, from France to Asia Minor, late May/June, the plants are remarkably hardy even in the clay soils of north-west England. While they will grow and flower well in a mixed border, bulbs may take several years to flower from seed; thenceforth clumps rapidly increase. HZ 2

N. siculum subsp. *bulgaricum* differs in having whitish flowers shaded pink, with green bases and veins. Found in the Balkans, Turkey and further eastwards, cultural requirements might be expected to be slightly drier than for *N. siculum* subsp. *siculum*.

SOURCES Illus: Rix & Phillips, *Bulb Book*; Synge & Hay, *Dictionary Garden Plants*.

A. sikkimense
Baker 1874

Here is an *Allium* stealing the deep clear blue properly reserved for a Gentian. Slight and dainty, flowering in two seasons from seed, this is an excellent plant for the rock garden.

Like many other luminous blue flowers

Allium sikkimense

found in the garden, *A. sikkimense* hails from central Nepal and western China, growing on open slopes at 3,000–4,800 m (9,000–14,000 ft). July.

In British gardens *A. sikkimense* flowers May–June differing from one of the other blue-flowered species, *A. beesianum*, by its earlier season and smaller size. By comparison *A. cyaneum* has long, prominent stamens and flowers July–August. *AA. kansuense, tibeticum* and *cyaneum* var. *brachystemon* have all been considered synonyms of *A. sikkimense*.

Slender stems, 15–25 cm (6–10 ins) carry lax umbels of 5–20 flowers on short stalks, with linear leaves equalling the length of the stems. Bell-like, blue flowers, nodding or erect, bear tepals 7 mm (⅓ ins) long. Most important details in identification are the blue stamens which are shorter than the perianth segments.

The species needs moisture but good drainage, flourishing outdoors. Slight and slender, the bulbs appear to be little more than fibrous roots; division in spring is easy. In cold, wet gardens, autumn division leaves unestablished plants at the mercy of slugs and damp. HZ 4

SOURCES Grey, *Hardy Bulbs*; *Eur Garden Fl*, 13; *Hortus III*. Illus: *Bot Mag*, 8858, 7290 (1893) *A. kansuense*.

A. sphaerocephalon
Roundheaded Leek
Linnaeus 1753

The ovoid bulbs have white or yellowish bulblets enclosed in the stem sheath. The reddish-purple umbels, 1–6 cm (⅖–2⅓ ins) are usually rather egg-shaped and occasionally carry bulbils. Below, the 2-valved spathe is persistent, 2 cm (¾ ins). Two–6 leaves sheathe half-way the hollow, hemicyclindrical stem, 5–90 cm (2–36 ins). The stamens are longer than the tepals. (Pl. p. 36.)

This readily obtainable *Allium* appears in many popular bulb catalogues and is an inexpensive plant for sunny borders in late July and August. Very easy to grow, requiring only good drainage, the neat heads bob in the breeze, attracting bees. HZ 1

Found on open ground from Europe, North Africa to West Asia, this is a variable plant with several subspecies.

A. s. subsp. sphaerocephalon (syns. A. descendens, A. s. var. typicum, A. s. var. descendens)
Has pink or reddish-purple flowers.

A. s. subsp. arvense (syns. A. arvense, A. s. var. viridialbum)
Has white tepals with a green or yellowish keel and smooth pedicels.

A. s. subsp. trachypus (syns. A. trachypus, A. arvense var. trachypus, A. s. var. trachypus)
Resembles *A. s.* subsp. *arvense*, except the pedicels are papillose.

A. loscosii (syn. A. purpureum)
This name has been given to variants with bulbils.

A particularly good form with large dark purple heads appeared in the local florist as a Dutch import, unfortunately without bulbs, and no seed was set.

A. sphaerocephalon is a rare native of the Avon Gorge, Bristol.

SYNONYM *A. descendens* L.

SOURCES *Fl Eur*, 90; *Fl Turkey*, 95; *Fl USSR*, 159. Illus: *Bot Mag*, 1764 (1815); Rix & Phillips, *Bulb Book*.

A. splendens (= glittering)
Miyama-rakkyo, Japanese
Willdenow ex Roemer &
Schultes 1830

A. splendens resembles *A. lineare* (qv) with brown, narrow bulbs on a short rhizome and a smooth, slightly ribbed stem, sheathed almost to its middle by shorter, linear leaves with rough margins. The rose tepals of the dense, many-flowered, hemispherical umbel are shorter than the stamens and have a prominent purple nerve. Differing from the very prominent capsule of *A. lineare*, that of *A. splendens* is shorter than the perianth segments. There are also differences in stamen formation.

Found growing in light woodland, scrub, stony slopes and meadows from eastern Siberia, North Korea and Japan, flowering in July–August. HZ 2

A. splendens thrives on well-drained, raised beds in our northern garden.

Miniature plants labelled *A. splendens* var. *kurilense* with enormously exserted stamens and styles, flowering in late autumn, are probably *A. thunbergii*.

SOURCES *Fl USSR*, 12; Ohwi, *Fl Japan*, 6.

A. stellatum
Ker-Gawler   1813

The outer coats of the ovoid bulbs are membraneous, producing a 30–70 cm (12–28 ins) stem, with 3–6 basal leaves, 1–2 mm wide and keeled. Hemispherical umbels are packed with upright, star-shaped flowers on slender, 1–2 cm (⅖–¾ ins) stalks. The stamens are longer or equal to the pink tepals with each capsule lobe bearing 2 crests.

A. s. 'Album'

Hortus III lists this as a variety.

Growing on rocky hills, prairies and arid slopes, from Ontario to Illinois, west to Saskatchewan, Wyoming and Oklahoma, *A. stellatum* is found flowering from July–September. This is a hardy *Allium* which does well in the open garden, being tall enough for a frontline situation in a well-drained, sunny border. HZ 2

Described by some as an upright form of *A. cernuum*, further confusion in gardens

has occurred by the possibility of keying plants as members of *A. senescens* group.

SOURCES *Eur Garden Fl*, 83; Gleason & Cronquist, *Manual of Vascular Plants of N Eastern U S & Adjacent Canada*. Illus: *Bot Mag*, 1576 (1813); Venning, *Wildflowers of N America*.

A. stellerianum
Willdenow   1799

This plant with yellowish-white, cup-shaped flowers on a short, 10–30 cm (4–12 ins), slightly ribbed stem, appears in *Flora Europaea* as a plant from the Central Urals. Narrow bulbs arise from a rhizome with thready basal leaves, the umbel is few-flowered, globular or hemispherical and both style and yellow-anthered stamens are exserted.

A degree of confusion arises over the occurrence of the spellings *A. stelleranum* and *A. stellerianum*. At one stage in nomenclature deliberations, it was decided that all names ending in -*er* should not be followed by an *i*. Since then the decision has been reversed, so now -*erianum* is correct.

That does not quite explain everything, for there are plants with different attributions. *Flora USSR* calls the plant *A. stellerianum* Willdenow, 1799, flowering July–August on slopes in Siberia and Mongolia, but also lists *A. stellerianum* Ledebour 1853 as a synonym of *A. rubens* (*Fl USSR*, 42). *Hortus III* refers to *A. stellaranum* Willdenow but the *RHS Dictionary* has an entry for *A. stellerianum* with yellowish or pink flowers, no attribution but a postscript, cf. *A. rubens*.

Grey has two entries for the same spelling: *A. stellerianum* Ledebour, a synonym for *A. flavidum*, and faint damning praise for *A. stellerianum* Willd.

SOURCES *Fl Eur*, 7; *Fl USSR*, 40; Grey, *Hardy Bulbs*.

A. stipitatum
(stipitate = with a little stalk)
Regel 1881

Blackish, papery tunics cover the 3–6 cm (1½–2⅓ ins) nearly spherical bulbs. Stems are ribbed and 60–150 cm (24–60 ins) tall, (described as smooth in *Flora USSR*). The 4–6 broad, grey-green, basal leaves are often densely hairy. (In *Flora USSR* and *European Garden Flora* they are described as hairy on the undersurfaces, while in *Flora Iranica* some forms may only be scabrid on the nerves underneath.) Attractive early in the year, like many 'drumsticks' they are withered by flowering time.

Umbels vary from hemispherical to spherical, with star-shaped, lilac-purple flowers held on almost equal pedicels, 2.5–5 cm (1–2 ins). The stamens are equal in length to the segments which twist and reflex after maturity. The nodular ovary has a short stalk. Occasional, less vigorous, white forms occur.

Grey in *Hardy Bulbs* describes plants distributed by Regel from St Petersburg (Leningrad) Botanic Garden, grown at Kew, with hollow stems, bright green leaves with a mixture of hairs, and sweetly scented flowers.

The confusion in description almost equals the mystification that bedevils the average gardener trying to sort out the various 'drumstick' alliums. A botanical merry-go-round of *AA. aflatunense, rosenbachianum* and *stipitatum* mocks nursery catalogues and buyers.

A. stipitatum comes from stony slopes around 2,500 m (6,000–7,000 ft), Afghanistan, Pakistan, the Pamirs and Tien Shan in central Asia, flowering May–June. Well-drained, sunny corners of the garden should suffice, for this is one of the easiest of its class and as a bonus numerous offsets are produced. HZ 4

SOURCES *Fl Iranica*, 121; *Fl USSR*, 207; *Eur Garden Fl*, 72. Illus: *Gartenflora* 30, 1881; Mathew, *Year Round Bulb Garden*.

A. suaveolens
(= fragrant)
von Jacquin 1789

A. suaveolens resembles the *A. senescens* group with slender bulbs on a rhizome and while like *A. angulosum* the linear, flat leaves are keeled on the undersurface, those of *A. suaveolens* sheathe the lower quarter of the 30–50 cm (12–20 ins) stem. Two persistent valves of a spathe underlie the hemispherical umbel with its cup-shaped flowers of pink or white with a pink keel. The long, white filaments 6–7 mm (⅖ in) capped with tawny anthers project well beyond the 4–5 mm (⅛ in) tepals, enhancing the size of the rather small 2–3.5 cm wide umbel (¾–1½ ins). In recompense, *A. suaveolens* is sweetly scented.

A common plant of damp meadows and moors in southern and central Europe, easily grown in the open garden. HZ 3

SOURCES *Fl Eur*, 10; Grey, *Hardy Bulbs*. Illus: Grey-Wilson, *Alpine Fl of Britain & Europe*.

A. subhirsutum
(= somewhat hairy)
Linnaeus 1753

The bulbs are spherical, membraneous-coated, with solid, 7–30 cm (3–12 ins) scapes carrying loose, hemispherical flower heads, 2.5–7 cm (1–2¾ ins) in diameter with a single-lobed spathe. Individual flowers, white, flat and star-shaped, hang on unequal stalks which can be up to 4 cm (1½ ins) long. The anthers are brown, occasionally yellow, much shorter than the perianth segments. The 2–3 linear, basal leaves are decidedly hairy, mainly along the margins.

A. subhirsutum is found flowering April–June in sandy or dry, stony places below 1,300 m (4,000 ft), throughout

A. subhirsutum (Author)

most of the Mediterranean. Plants are raised easily from seed and, while requiring summer rest, can be grown in warm sunny spots, otherwise frame or pot culture. In suitable sites *A. subhirsutum* makes a good garden plant; the flowers are pleasantly if faintly scented. HZ 5

A. permixtum found in the central Mediterranean is very similar, lacking only the hairy leaves.

SYNONYMS *A. clusianum, A. ciliatum, A. hirsutum, A. niveum.*

SOURCES *Fl Eur*, 30; *Fl Turkey*, 8; *Eur Garden Fl*, 32. Illus: Rix & Phillips, *Bulb Book*; Polunin, *Fl of Greece & Balkans*.

A. subvillosum
(= somewhat shaggy)
Salzmann ex Schultes &
Schultes 1830

A. subvillosum resembles *A. subhirsutum* (qv) but its white flowers are a little more cup-shaped. The leaves are also hairy along the margins but the spathe may split into 3–4 lobes. There are rather more flowers in the 2.5–5 cm wide umbel (1–2 ins).

Flowering March–May on sandy beaches and fields in Greece, Sicily, the western Mediterranean through to Portugal, the species is also very common in North Africa. HZ 5

SYNONYM *A. subhirsutum* var. *subvillosum.*

SOURCES *Fl Eur*, 32; *Eur Garden Fl*, 33; Grey, *Hardy Bulbs*. Illus: Redouté, *Liliacées*, 6: t.305 (1812).

A. textile (textilis = woven)
Prairie Onion
Nelson & Macbride 1913

Several clustered bulbs with reticulated, grey coats, themselves enclosing 1 or more bulbs, produce 2 leaves which while green at flowering are still present as seeds form. They are about the same length as the 6–25 cm (2–10 ins) stem. Below the umbel (15–30+ flowers) is a 3-valved spathe. The curving flower stems, unequal and 2–3 times the length of the tepals, become rigid as the seeds form. White or occasionally pink tepals, 5–7 mm (⅓ ins) long, have a reddish-brown mid-rib. While the outer tepals are erect, the inner are narrower with the tips flattening outwards

Allium textile

slightly, with stamens shorter than the tepals. The ovary has 6 obvious knobs on top of the lobes.

A. textile is one of the commonest alliums of the northern prairies, found throughout the dry plains and hills from Minnesota and Manitoba to Alberta, south to Colorado, northern Mexico and Utah. Flowering in May–June, the species can often be found in thickly carpeted colonies.

A dry summer rest is usually necessary to prevent the bulbs rotting; bulb frame or pot culture may be required. HZ 2

SYNONYMS *A. reticulatum, A. r.* var. *playanum, A. aridum.*

SOURCES *Hortus III.* Illus: Hitchcock *et al., Vasc Pl Pac N W*; Cronquist *et al, Intermountain Flora*; Young & Young, *Colorado West*; Niehaus & Ripper, *Field Guide S Western & Texas Wild Flowers.*

A. thunbergii
Don 1827

Several attractive small alliums are becoming available to gardeners from the Far East, particularly Japan. Difficulties abound however when attempting to authenticate this group, accounts in English being hard to find. The majority of the synonyms listed appear in the index to the *Fl Rei Pop Sin. A. thunbergii* fortunately is reasonably well documented.

Clumps of narrow bulbs with fibrous roots and no obvious rhizome produce 30–60 cm (12–24 ins) stems; most plants in cultivation however average 15–20 cm (6–8 ins). The softly 3-sided, erect leaves are somewhat shorter but unwithered at flowering time. Long pedicels, 1–1.5 cm (⅖–⅗ ins), carry a full head of rose-purple flowers, the most prominent feature being the enormously elongated stamens and style which effectively doubles the width of the umbel, up to 3 cm (1¼ ins).

A further bonus to this charming small

Allium thunbergii

plant is the late flowering season, September to November.

Growing on low mountains in Honshu, Shikoku, Kyushu and extending from Japan into South Korea, *A. thunbergii* should be inured to cold and damp but the late flowering season makes pot culture more appropriate if the grower wishes to enjoy the flowers unspoiled by weather. HZ 4

A. t. 'Ozawa'

This selection was introduced into the USA by George Schenk and named after the original selector, for which 'Ozoke' appears to be a misprint.

SYNONYMS *A. japonicum, A. pseudojaponicum, A. sacculiferum, A. taquetii, A. ophiopogon, A. morrisonense, A. pseudocyaneum, A. komarovianum.*

SOURCES Ohwi, *Fl Japan*, 7; *Fl Rei Pop Sin*, 84; *Hortus III*. Illus: *Bulletin AGS*, Vol. 52, p. 353.

A. tibeticum
Rendle 1906

Confusion has arisen over plants available in horticulture labelled *A.* sp. Tibet, which became *A. tibeticum*, itself usually a misnomer for *A. cyathophorum* var. *farreri* (qv), a plant with reddish-purple flowers.

The species with blue tepals described by Grey from Tibet around Lhasa, to 4,875 m (10,000 ft), is now considered a synonym of *A. sikkimense*.

SOURCES Stearn, '*Allium & Milula* in C & E Himalaya', 9; *Eur Garden Fl*, 13.

A. tricoccum
Wild Leek, Ramps
Aiton 1789

A. tricoccum can be compared to two other broad-leaved alliums, *A. ursinum* and *A. victorialis*, having more in common with the former. Its common name of Ramps is also similar to that of Ramsons, used for *A. ursinum*. A native of the eastern parts of the USA, two races have been described. The first can be found in damp soils in rich woodland, flowering one to two weeks before plants of the second race, *A. tricoccum* var. *burdickii*. This latter frequents woods in the uplands and has a greener stem with paler green, more glaucous leaves than the type.

The bulbs are slender, ovoid, with netted coats, attched to a short rhizome. The 2–3 bright green, shining, broad leaves, 10–30 cm × 2.5–5 cm (4–12 ins × 1–2 ins), appear in early spring but have withered before the flowering stem appears, unlike the pattern of *A. ursinum*, its Old World counterpart.

Often found in large colonies, Ramps are gathered in Appalachia early in the year, the focus for 'ramp-eating festivals'. Mountain dwellers in west Virginia are reported as being especially fond of the flavour, the strong-scented plants being eaten in place of Spring Onions.

By midsummer the purplish scape, 10–40 cm (4–16 ins), carries the loose, many-flowered, hemispherical umbel. Star-shaped flowers, with white tepals 4–6 mm (¼ ins), hang on stalks 1–2 cm (⅖–¾ ins) long.

In his *Flora of the Southeastern United States* (1903) and his *Manual of the South-eastern Flora*, Small treated *A. tricoccum* as a separate genus with a single species, *Validallium tricoccum*.

Flowering in late June to early July, from Georgia to Minnesota and New Brunswick, *A. tricoccum* has been used for naturalising in woodland. The species is also found in parts of the Himalaya around 2,000–5,200 m (6,500–17,500 ft), in moist meadows and scrub. Care should be taken that plants do not overrun their allotted space (Fig. p. 25). HZ 2

SOURCES Gleason & Cronquist, *Manual Vascular Plants of N Eastern US & Adjacent Canada*; *Eur Garden Fl*, 23; Stearn, '*Allium & Milula* in C & E Himalaya'. Illus: Moore in *Baileya*, Vol. 2, 1954, Fig. 34; Rickett, *Wild Fl U S*, Vol. 1; Stupka, *Wildflowers in Colour*.

A. trifoliatum
(= three-leaved)
Cyrillo 1792

Resembling *A. subhirsutum* (qv), *A. trifoliatum* differs in the following particulars; the leaves are less hairy, the shuttlecock umbel carries stellate flowers with tepals 7–10 mm (⅖ in), white with a pink stripe or suffused with the palest pink, on shorter pedicels, 1½–3 times the length of the individual segments. The flowers may deepen in colour as they age.

Flowers are found on stony or cultivated land below 500 m (1,500 ft), throughout the Mediterranean from Crete to the Riviera. Flowering March to May, the long, tapering leaves may still be green at anthesis. HZ 5

SYNONYMS *A. graecum*, *A. subhirsutum* subsp. *trifoliatum*, *A. s.* var. *graecum*, *A. s.* var. *trifoliatum*.

SOURCES *Fl Eur*, 31; *Eur Garden Fl*, 34; *Fl. Turkey*, 9. Illus: Rix & Phillips, *Bulb Book*.

A. triquetrum
Linnaeus 1753

A native of damp, shady places and woods throughout the western Mediterranean, the species has become naturalised in Britain, flowering in spring.

The basal, rather fleshy leaves are keeled, while the stem is distinctly 3-sided (triquetrous – from which this species derives its name). The loose umbel, 4–7 cm (1–2¾ ins) wide, carries 3–15 flowers on longish flower stalks, usually held 1-sidedly, reminiscent of the growth form of the English Bluebell (*Hyacinthoides non-scripta*). The flowers are white with a green stripe and bell-shaped, 1–1.9 cm × 2–5 mm (¼–¾ ins × ¹⁄₁₀–⅕ ins), effective plants for growing in the shade of shrubs or trees, although invasive when well suited. In north-west England, drier conditions are required than the streamsides of their Mediterranean origins (Fig. p. 27.) HZ 4

Seed dispersal is effected by ants. After fertilisation, the stem wilts and the flower head carrying the ripening seed rests on the ground. Aril, which is an accessory seed covering, attracts the insects which then carry the seeds away to scatter over the surrounding ground.

One of the alliums approved by Farrer, he described the flowers thus: 'of a diaphanous white, looking like the ghost of a dead white flower drowned long ago in deep water.'

SOURCES *Fl Eur*, 36; *Fl Turkey*, 17; *Hortus III*. Illus: Rix & Phillips, *Bulb Book*; Grey-Wilson & Mathew, *Bulbs*; Polunin, *Concise Fl Europe*.

A. tuberosum
Chinese Chives, Kiu ts'ai
Rottler ex Sprengler 1825

The similarities and differences between *A. tuberosum* and *A. ramosum* have been listed under the entries for the latter.

A. *tuberosum* has conical bulbs on a short rhizome, a solid stem, 25–50 cm (10–20 ins), sheathed to one-eighth its length by 4–9 solid leaves with a slight keel to the undersurfaces. Numerous white, starry flowers, on 1–3 cm (²⁄₅–1 ¹⁄₅ ins) pedicels, pack the usually hemispherical umbels. The 4–7 mm segments are white with a faint green or brown line on the undersurfaces, the stamens are only slightly shorter and the capsule is widest above its middle.

A very easy, attractive plant for the open garden, willing even to grow in shade in Lancashire, flowers continue to open in sequence from August into October. The leaves remain green and pleasing at anthesis, the plants steadily increasing in clumps. Further bonuses are the fragrance and edibility of both the flowers, which can be used in salads, and the leaves. HZ 3

A native of eastern Asia it has been used as a salad crop for centuries in China, Japan, South-East Asia and through to India. The Chinese characters forming the name are possibly those of a vegetable listed in a Dictionary of the Han period (c.206BC–AD220).

A notion of the confusion surrounding the identities of *A. tuberosum* and *A. ramosum* is outlined in the entries for the latter. Many synonyms have accrued to the title of *A. tuberosum*—see Stearn, 'Nomenclature & Synonymy of *Allium odorum* & *Allium tuberosum*', *Herbertia*, 1944.

SYNONYMS *A. uliginosum*, *A. tuberosum* Roxburgh.

SOURCES Stearn, *Allium* & *Milula* in C & E Himalaya, 10; *Fl Rei Pop Sin*, 25, Illus: *Bot Mag*, new series, 386, 1962; Rix & Phillips, *Bulb Book*.

A. unifolium One-leaved Onion
Kellogg 1863

Despite its common name, this is not a one-leaved Onion! Allowing for this misnaming, *A. unifolium* is possible the most attractive of all the American onions. Flowering in May–July in moist soils on 170–1,700 m (500–5,000 ft) slopes, in the Coast Ranges of California into Oregon, it is only marginally hardy in north-west England.

The bulbs are unusual, produced on short, lateral rhizomes from the outside of the previous bulb, only the root pad of the parent has remained by the time seed has formed. The 2–3 flattish leaves are shorter than the rather stout stem, 20–60 cm (8–24 ins), while the many-flowered umbel carries the loveliest flowers on long pedicels, 2–2.5 cm (³⁄₄–1 ins). The tepals are a

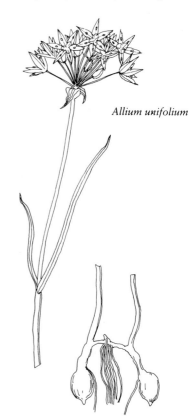

Allium unifolium

A. *unifolium* (Author)

bright sugar pink, glistening like icing, sometimes with a deeper mid-vein, the whole a flaring bell shape and with age becoming papery. Occasionally plants with white tepals are found, equally lovely. Pale pink filaments have deeper anthers, rose, mauvish or tinged with brown pollen, contrasting with the tepals.

While *A. unifolium* is a little tall for a rock garden and will need a summer rest, it is a lovely plant, also well worth growing in a bulb frame or as a pot plant. HZ 4

A. murrayanum (qv) of gardens appears to belong to the same species.

SYNONYMS *A. grandisceptrum, A. unifolium lacteum.*

SOURCES Munz, *California Flora*; Grey, *Hardy Bulbs.* Illus: Rix & Phillips, *Bulb Book*; Hitchcock *et al., Vasc Pl Pac N W*; Abrams, *Illus Fl Pac States.*

A. ursinum
(ursa = bear) Ramsons
Linnaeus 1753

This is the Wood Garlic that fills damp woods in spring with a powerful onion smell. The 'perfume' is unfortunate for this is a very lovely flower, the leaves are handsome and plants that will grow in the shade that Ramsons enjoy are valuable denizens of woodland. This is also an onion with a reasonable flavour, one manner in which unwelcome spread might be limited.

The edible bulbs are papery and cylindrical. The 2–3 leaves are broadly elliptical, dark green, with a narrow basal stalk, 6–20 cm × 1.5–8 cm (2¼–8 × ¾–3¼ ins) and appear very similar to Lily-of-the-Valley (*Convallaria majalis*). Simply handling the leaf makes the difference easy for the nose to judge. The leaves have turned themselves upside down; having the majority of their stomata, which normally occupy the undersurface of leaves, reversed and appearing on the topside. The stem is 2–3-angled, 10–20 cm (4–8 ins). Above the 2 bracts, the flower head is quite large, 2.5–6 cm (1–2½ ins) in diameter, with 5–20 flowers. The star-shaped tepals are a lovely glistening white with prominent ovules.

A spring flower in woods throughout Europe and the European parts of the USSR, culture creates little difficulty, indeed invasive might be the appropriate adjective (Fig. p. 18). HZ 1

Two subspecies are described, *ursinum* and *ucrainicum*, which differ in the roughness of the pedicels.

Dispersal of the fertilised seed is by ants. The stem flops on the ground as the seed matures. The seed coat, impregnated with an aromatic oil, attracts the insects which carry their booty away, dropping some in their travels. The scattered seed, having found fresh areas to colonise, does the job most thoroughly.

The New World has a comparable species—*A. tricoccum* (qv).

SOURCES *Fl Eur*, 39; *Fl USSR*, 2; *Hortus III*. Illus: Rix & Phillips, *Bulb Book*; Polunin, *Concise Flora Europe*.

A. validum
(validus = robust growing)
Tall Swamp Onion
S. Watson   1871

To find an Onion growing in a swampy region is really quite extraordinary in a genus more partial to baking in deserts or drying on a mountainside. *A. validum* is odd rather than beautiful, the umbel rather too small for the length of stem. In a boggy garden *A. validum* would make an interesting clump but growing in a dry border plants look dingy, dowdy and dwarfed. However, if the plants do not please they can be eaten, the bulbs having a good flavour.

The ovoid bulbs rise from a thick, Iris-like rhizome covered with coarse, parallel fibres, topped by an angled, flattened stem, 30–80 cm (12–30 ins), with 3–6 slightly shorter, ridged or flat leaves. These are still green at flowering time. Below the hemispherical umbel, containing 15–30 flowers, the spathe splits into 2–4 lobes. Flower colour varies from deep rose to nearly white, with yellow or purplish stamens much longer than the 5–10 mm segments. The tepals wither and the flower stalks, 10–18 mm (²⁄₅–⁴⁄₅ ins) in length, elongate as seed forms, then become rigid.

In swampy meadows, 2,500–3,000 m (7,000–9,000 ft) mainly in the Coast Ranges, Cascades and Sierras, *A. validum* flowers June to August. Plants can only be

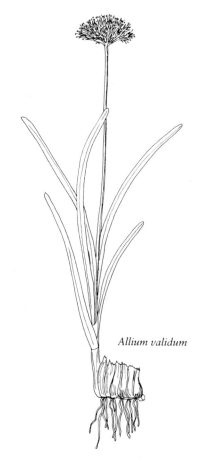

Allium validum

confused with *A. brevistylum*, which is much smaller and whose stamens are shorter than the tepals. HZ 2

SOURCES Munz, *California Flora*, 2; Ownbey, 'Genus *Allium* in Idaho'. Illus: Hitchcock *et al.*, *Vasc Pl Pac N W*; Cronquist *et al.*, *Intermountain Flora*, 6; Niehaus & Ripper, *Field Guide to Pacific State Wildflowers*.

A. vancouverense
Macoun 1888

Described by John Macoun from the summit of Mount Arrowsmith, Vancouver Island, Canada, this is now considered a synonym of *A. crenulatum*.

A. victorialis
Linnaeus 1753

The botanical name for this species is derived from medieval European mythology. *Siegwurz* in German means Victory root, alluding to the belief that miners held of its ability to protect them against evil spirits that roved underground. As pungent as Garlic, both the odours of damp earth and the malice of predatory ghouls should be vanquished by the smell.

Cylindrical bulbs with densely fibrous coats gather around a short rhizome. Two or 3 broad, elliptical leaves, 8–25 cm × 1.5–9 cm (3¼–10 × ¾–3½ ins), have a short stalk with purple-tinged sheaths covering the flower stem for almost half its length. The tips of the leaves gradually taper, the whole blade tending to fold longitudinally, particularly when young. The 30–60 cm (12–24 ins) stem is 2-angled at the base. Unfortunately the flower heads are small for the size of the plant, 3–5 cm (1¼–2 ins), spherical or hemispherical with numerous, star-shaped flowers. The greenish-white tepals are markedly shorter than the stamens, which have yellow anthers (Fig. p. 17).

A. victorialis var. listera

This is included in *Flora Rei Pop Sin*, 1.

Found throughout southern Europe and northern Asia from the Caucasus to Siberia, the Himalayas to Japan, several regional variations are listed in *Fl USSR*. *A. victorialis* flowers on mountain pastures and stony slopes, where its clumps may be seen in June, July. While not a common plant, in favoured localities it may form large colonies. Though not often found in cultivation, a sunny border and good drainage should suffice. HZ 3

Protected in some countries, *A. victorialis* has been valued as a prophylaxis against scurvy. Widespread use has been made of its food potential, both fresh and pickled. The plants were salted in bulk for winter use in many areas of Siberia.

SYNONYMS *A. microdictyum, A. ochotense, A. latissimum.*

SOURCES *Fl Eur*, 22; *Fl USSR*, 1; *Hortus III*. Illus: *Bot Mag*, 1222 (1809); *Macdonald Encyclopedia of Alpine Flowers* (1984).

A. vineale Crow Garlic
Linnaeus 1753

A pestilential weed included as a warning. Tall stems, 35–100 cm (14–40 ins) in height, sheathed on the lower half with hollow, cylindrical leaves, carry 2–4 cm (¾–1½ ins), almost round umbels usually packed with bulbils. The spathe is single-valved, 3 cm (1¼ ins) or longer, with a beak as long as the base. Occasional forms carry only bell-shaped flowers which can be greenish, reddish or pinkish, with yellow anthers more or less projecting, on 5–30 mm (¼–1¼ ins) pedicels. More often the head is a mixture of flowers and bulbils, each one of which rolls away to multiply and flourish (Fig. p. 59).

Several varieties—*compactum*, *typi-*

cum, *capsuliferum*, *purpureum* and *virens*—have been described.

Similar plant thugs are *A. scorodoprasum* with flat leaves and a single-valved bract, 1.5 cm (3/5 ins), short-beaked and deciduous, and *A. carinatum* with 2 unequal prominent valves forming the spathe.

Found over most of Europe, Asia Minor, North Africa and West Asia, plants spread to America, probably in colonial times. The bulbils may contaminate wheat fields, finding their way into the grain harvest and spoiling the flour. Few would relish garlic bread with marmalade or honey. Similarly, infested pastureland can flavour cows' milk.

SYNONYMS *A. compactum*, *A. rilaense*, *A. kochii*, *A. affine*, *A. assimile*.

SOURCES Illus: Grey-Wilson & Mathew, *Bulbs*; Fitter & Fitter, *Wild Flowers of Britain & N Europe*; Grey-Wilson, *Alpine Flowers of Britain & Europe*; Huxley, *Mountain Flowers in Colour*; Hitchcock *et al.*, *Vasc Pl Pac N W*.

A. virgunculae
(virguncula = young girl)
Maekawa & Kitamura   1952

In the comparatively short time since this small Japanese *Allium* has been described, it has become quite widely distributed and known to *Allium* enthusiasts, receiving an Award of Merit in 1983. Part of its popularity rests on the late-flowering period—November, and the almost perennial quality of the grass-like foliage which, unlike so many onions, is green and fresh during flowering.

Narrow, clustered, slender bulbs with pale rose-coloured, membraneous coats produce about 3–5 dark green, linear leaves, 10–20 cm (4–8 in), dotted with tiny, white spots. Slender stems, 8–22 cm (3–9 ins), bear 2–12 pink, star-shaped flowers in a loose umbel backed by

a single-lobed spathe. The green banded tepals are 5 mm × 3–4 mm (1/5 × 1/5 in) with longer stamens, 6 mm (1/4 in); the style is also exserted. A white form has been recorded.

A. virgunculae grows only on Hirato Island, Kyushu and while it has been successful outside in a sunny site on the rock garden at the Royal Botanic Gardens, Kew, most growers may feel the size and late flowering season suggest pot culture.

Being almost perennial, plants can be divided during winter or spring. Dainty and quite easy to grow, *A. virgunculae* is a charming subject for the alpine house. HZ 3

SOURCES Ohwi, *Fl Japan*, 9; *Eur Garden Fl*, 47; Mathew, *Smaller Bulbs*. Illus: *Bulletin AGS*, Vol. 51, p. 294.

A. viviparum
Karelin and Kirilow   1841

A synonym for a form of *A. coeruleum* (*Flora USSR* spelling), and a very polite way of describing a flowering head full of bulbils. Any flowers in the umbel are blue and Grey suggests growing on seedlings for colour, while roguing out bulbilliferous forms.

See *A. caeruleum*.

Hortus III lists the plant as *A. caeruleum* var. *bulbilliferum*. South Siberia.

SOURCES *Fl USSR*, 129; Grey, *Hardy Bulbs*.

A. wallichii
Kunth   1843

A maroon-purple *Allium*, that is found in Nepal in those areas where the monsoon rains drench its habitat, should be an excellent addition to wet Western gardens in August and September.

Plants are named after Dr Nathaniel Wallich, who was Superintendent of the Calcutta Botanical Garden.

A 3-angled flowering stem, 30–90 cm (12–36 ins), carries a loose umbel, 5–7 cm (2–3 ins) wide, with numerous long-stalked florets. The papery tepals are longer than the purple stamens and ovary, the whole flower being star-shaped. The numerous leaves are keeled, 2 cm (¾ ins) broad, sheathing the lower stem almost all at the same level. Numerous fibrous roots are present in place of bulbs.

A. *wallichii* is found growing in western China to Nepal and Pakistan at 2,800–4,300 m (8,000–12,000 ft), in clearings between trees or shrubs. HZ 2

A. *wallichii* var. *platyphyllum* (Diels) J. M. Xu (syn. A. *polyastrum* var. *platyphyllum*, A. *platyphyllum* (Diels) Wang & Tang, A. *lancifolium* Stearn)

Appears in *Flora Reipublicae Popularis Sinicae*, the text being in Chinese. An English key in preparation by Dr Peter Hanelt and colleagues should help clarify any differences between A. *wallichii* and A. *polyastrum*.

This is an easy plant to grow in a border, most welcome at the latter end of summer. Seed is formed very quickly after flowering, but plants have not proved invasive in north-west-England. Clumps are easily propagated by division in spring, though the shoots are slow to come through the soil and may easily be dug up by accident.

SYNONYMS A. *tchongchanense*, A. *polyastrum*, A. *bulleyanum*.

SOURCES *Eur Garden Fl*, 14; *Fl Rei Pop Sin*, 10, 10a. Illus: Polunin, *Flowers of the Himalaya*.

A. *yunnanense* Diels 1912

An account of the confusion surrounding AA. *yunnanense*, *mairei*, *amabile*, *pyrrorhizum* and *acidoides* is given under A. *mairei*. Plants were collected by George Forrest in south-west China.

An *Allium* suitable for the rock garden or a raised bed and flowering in late summer, this small plant requires moisture and good drainage. HZ 5

See A. *amabile* and A. *mairei*.

A. *zebdanense* Boissier & Noë 1859

Originally collected in Lebanon, this is one of the most attractive of the white alliums. E.B. Anderson in *Rock Gardens* wrote that it might be invasive, while listing A. *cyathophorum* var. *farreri*, a notorious seeder, without making any comment, suggesting his experience may have differed from that of many other gardeners. In fact A. *zebdanense* seems to require an extremely dry, sharply drained situation for its survival in the open garden and will flourish rather better in a bulb frame or alpine house. Plants have entered dormancy by June, requiring protection from rain.

Seed is offered in several lists but I have never had successful germination. Bulbs are easily available commercially.

Grown as an attractive pot plant, the observer will be pleased to notice an attractive faint perfume that belies the odoriferous reputation of the onion tribe. April–May. HZ 4

The loose umbel carries 6–10 large, starry flowers with glistening white tepals, 9–13 mm (¼–½ ins), blunt ended and longer than the stamens, on unequal pedicels. The 2–3 leaves are slightly shorter than the 25–40 cm (10–16 ins) stem. The bulbs are small and ovoid with brown coverings. (Pl. p. 11.)

SYNONYMS A. *chionanthum*, A. *candolleanum*.

SOURCES *Fl Turkey*, 13; *Eur Garden Fl*, 29; Grey, *Hardy Bulbs*. Illus: Rix & Phillips, *Bulb Book*.

APPENDIX I *Section Definition According to Flora Europaea (1978)*

The following is a simplified scheme of species classification into Sections.

Section RHIZIRIDEUM

Bulbs conical bulb on short rhizome
Leaves almost basal or sheathing lower half of the stem, flat not hollow
Stem angled or cylindrical
Spathe usually shorter than flower stems
Floret usually cup-shaped or bell-shaped, rarely star-shaped or almost globular
Stamens simple
Ovary with nectaries
Ovules 2 in each compartment
Stigma entire or 3-lobed
Seeds angular

1 *angulosum*	10 *suaveolens*
2 *rubens*	11 *lineare*
3 *senescens*	12 *palentinum*
4 *saxatile*	13 *hymenor-*
5 *horvatii*	*rhizum*
6 *albidum*	14 *obliquum*
7 *stelleranum*	15 *narcissiflorum*
8 *ericetorum*	16 *insubricum*
9 *kermesinum*	17 *inderiense*

Section SCHOENOPRASUM

Bulbs narrow on a short rhizome
Stems circular
Leaves almost basal or sheathing the lower quarter to one-third of the stem, cylindrical, hollow
Spathe equalling or shorter than flower stalks
Floret bell-shaped
Stamens simple
Ovary with deep nectaries
Ovules 2 in each compartment
Stigma entire
Seeds angular

18 *schoenoprasum*   19 *schmitzii*

Section CEPA

Bulbs approximately cylindrical, usually clustered on a short rhizome
Leaves sheathing lower part of stem, arranged in 2 vertical rows, hollow
Stem cylindrical, hollow
Spathe shorter than or almost equalling flower stalks
Floret star- or bell-shaped
Stamens simple or with small teeth at base of inner filaments
Ovary with distinct nectaries
Ovules 2 in each compartment
Stigma entire
Seeds angular

20 *cepa*   21 *fistulosum*

Section ANGUINUM

Bulbs Bulbs almost cylindrical, clustered on a short rhizome, with densely netted, fibrous coats
Stem 2-edged at the lower end
Leaves sheathing lower third-half of the stem
Spathe shorter than flower stalks, 1–2-valved, persistent
Floret star- or cup-shaped
Stamens simple

Ovary with distinct nectaries
Ovules 2 in each compartment
Stigma simple
Seeds almost globular

22 *victorialis*

Section MOLIUM

Bulbs ovoid or almost globular, no rhizome
Leaves almost basal with short sheath above the ground, flat
Stem cylindrical or angled
Spathe shorter than or equalling flower stalks
Floret star- or bell-shaped, or cylindrical
Ovary with distinct nectaries
Ovules 2 in each compartment
Stigma entire
Seeds angular

23 *roseum*
24 *circinnatum*
25 *massaessylum*
26 *phthioticum*
27 *breviradium*
28 *neapolitanum*
29 *longanum*
30 *subhirsutum*
31 *trifoliatum*
32 *subvillosum*
33 *moly*
34 *scorzoneri-folium*

Section BRISEIS

Bulbs semiglobular, no rhizome
Leaves almost basal with short above-ground sheath, so strongly keeled as to be 3-cornered
Stem 3-cornered, flaccid after flowering
Spathe shorter than the flower stalks, 2-valved, persistent
Floret bell- or star-shaped
Stamens simple, shorter than the tepals
Ovary with minute nectaries
Ovules 2 in each compartment
Stigma 3-lobed
Seeds angular

35 *triquetrum*
36 *pendulinum*
37 *paradoxum*

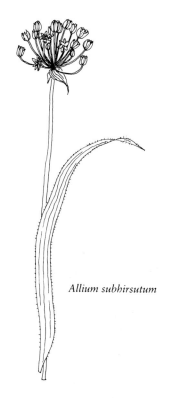

Allium subhirsutum

Section CHAMAEPRASON

Bulbs semiglobular, no rhizome
Leaves basal
Stem very short, umbel almost stemless in the rosette of leaves
Spathe shorter than the flower stems which are recurved in fruit
Ovary with 2 ovules in each compartment
Stigma entire
Seeds angular

38 *chamaemoly*

Section OPHIOSCORODON

Bulbs cylindrical not rhizomatous
Leaves basal, stalked, with broad blade
Stem sharply angled
Spathe shorter than flower stalks, 2-valved, persistent
Floret star-shaped

Stamens simple
Ovary 2 ovules in each compartment
Stigma simple
Seeds semiglobular

39 *ursinum*

Section SCORODON

Bulbs ovoid or semiglobular, not rhizomatous
Stem cylindrical
Leaves sheathing lower ¼–½ of stem, usually threadlike
Spathe usually shorter than the flower stalks
Flower tubular to star-shaped
Stamens simple or the inner sometimes with 2 short, basal teeth
Ovary distinct nectaries
Ovules 2 per compartment
Stigma entire
Seeds angular

40 *moschatum*	48 *grosii*
41 *inequale*	49 *cupanii*
42 *bornmuelleri*	50 *callimischon*
43 *rubellum*	51 *obtusiflorum*
44 *meteoricum*	52 *parciflorum*
45 *delicatulum*	53 *rouyi*
46 *frigidum*	54 *caeruleum*
47 *chrysonemum*	55 *sabulosum*

Section CODONOPRASUM

Bulbs ovoid, no rhizome
Stem cylindrical
Leaves sheathing up to two-thirds of the stem
Spathe 2-valved, the valves unequal, each a tail, longer than the flower stalks
Floret cup-, bell-, funnel-, cylinder- or egg-shaped, never star-shaped
Stamens simple
Ovary with or without minute, inconspicuous nectaries
Ovules 2 per compartment

Stigma entire
Seeds angular

56 *paniculatum*	66 *pilosum*
57 *pallens*	67 *luteolum*
58 *sipyleum*	68 *flavum*
59 *rupestre*	69 *melanantherum*
60 *favosum*	
61 *macedonicum*	70 *tardens*
62 *podolicum*	71 *carinatum*
63 *oleraceum*	72 *hirtovaginum*
64 *parnassicum*	73 *hymettium*
65 *staticiforme*	74 *stamineum*

Section ALLIUM

Bulbs ovoid or subglobose, no rhizome
Stem usually cylindrical
Leaves linear, flat or hollow, sheathing the lower quarter or more of the stem
Spathe 1- or 2-valved, usually beaked
Floret cylindrical, bell-shaped or oval; with permanently converging bases to the tepals
Stamens outer stamens usually differing in shape from the inner ones
Ovary distinct nectaries
Ovules 2 per compartment
Stigma entire
Seeds angular

75 *sativum*	91 *proponticum*
76 *ampeloprasum*	92 *melananthum*
77 *polyanthum*	93 *pruinatum*
78 *scaberrimum*	94 *regelianum*
79 *atroviolaceum*	95 *vineale*
80 *bourgeaui*	96 *amethystinum*
81 *commutatum*	97 *guttatum*
82 *pardoi*	98 *dilatatum*
83 *talijevii*	99 *gomphrenoides*
84 *baeticum*	
85 *acutiflorum*	100 *jubatum*
86 *pyrenaicum*	101 *rubrovittatum*
87 *scorodoprasum*	102 *integerrimum*
88 *albiflorum*	103 *chamaespathum*
89 *pervestitum*	104 *heldreichii*
90 *sphaerocephalon*	

A. rubrovittatum (Author)

Section MELANOCROMMYUM

Bulbs semiglobular or oval, no rhizome
Leaves basal with no above-ground sheath
Stem usually longer than leaves, cylindrical
Spathe shorter than flower stalk, becoming split into 2–4 remnants, persistent
Floret usually star-shaped, with tepals bending backwards as the flower ages
Stamens simple
Ovary 4–8 ovules in each compartment
Stigma entire
Seeds angular

Section KALOPRASUM

Bulbs semiglobular, not rhizomatous
Stem about as long as leaves, cylindrical
Leaves basal, no above-ground sheath
Spathe shorter than flower stalks, 2-valved, persistent
Floret bell- or star-shaped, tepals erect or bending backwards as the flower matures
Stamens simple, longer than the tepals
Ovary 6–14 ovules per compartment
Stigma entire
Seeds angular

110 *caspium*

105 *atropur-*
     pureum
106 *nigrum*

107 *cyrilli*
108 *decipiens*
109 *orientale*

APPENDIX II *Allium Collections*

Very few centres in the Western world carry comprehensive collections of *Allium*. The Royal Botanic Gardens, Kew in all probability carry the largest concentration world-wide. Approximately 225 species, subspecies and varieties were listed in 1988.

The Catalogue of Plants in the Royal Botanic Garden, Edinburgh in 1986 contained around 65 entries. Cambridge Botanic Garden displays wild-collected specimens in the rock garden area, as well as a separate genus bed.

A. cyaneum (John Fielding)

Several of the autumn-flowering Asiatic *Allium* can be seen at Threave Gardens, Castle Douglas, Galloway, Scotland; attractive plantings of *AA. beesianum, cyaneum, polyastrum* and *tuberosum* are flowering in early September.

Several of the suppliers have displays in borders or raised beds. Bressingham Gardens have several large plantings of *AA. senescens, schoenoprasum* 'Forescate', *nuttallii, cernuum* and *cyaneum* in the Dell Garden—showing how well alliums can blend with other flowers.

A. christophii and *A. aflatunense* have been used effectively in the Laburnum pergola at Barnsley House, and in the dry garden at Kiftsgate.

The National Collection of *Allium* is held for the National Society for the Preservation of Plants and Gardens by Pat Davies, 6 Blenheim Road, Caversham, Reading, Berks RG4 7RS.

A taxonomic study is being conducted by the Akademie der Wissenschaften der DDR, Zentralinstitut fur Genetik und Kulturpflanzenforschung in Gatersleben, Germany under the direction of Dr Peter Hanelt.

In America, Denver Botanic Garden has a collection of native and foreign *Allium* in the Rock Garden.

Berkeley University Botanic Garden, California has excellent specimens in the Native Plants Collection. These are well labelled with their provenance.

American *Allium* are best studied, alongside other genera of the suberb and largely neglected native flora, in their natural environment. Field trips of the local Chapters of the American Rock Garden Society are highly recommended. Plants abound in the numerous National Parks.

APPENDIX III *Societies and Suppliers*

SOCIETIES

The seed lists of the following societies provide a variable number of species. There is, however, no guarantee that the naming is accurate. Unfortunately, plants circulate in good faith with wrong labels and this error may be perpetuated from grower to grower. Seed in many exchanges will be pooled, collections of any one species sent in from differing sources will be mixed before individual packages are made up. Some societies now attach donor names to items on the list.

Seed obtained from botanic gardens may also carry erroneous labels. If a botanist on the garden staff has a particular interest in a genus then the relevant seed is more likely to be correct.

Commercial seed merchants are required to make more stringent checks on their produce. The passage of Trade Descriptions Acts have had some influence even in products not legally bound by such legislation. Difficulties may still arise where the taxonomic accuracy of a species is unclear.

Alpine Garden Society
Secretary, AGS Centre, Avon Bank, Pershore, Worcestershire WR10 3JP

American Rock Garden Society
Membership Secretary, Carole Wilder, 221 West 9th Street, Hastings, MN 55033, USA

Hardy Plant Society
General Secretary: Tricia King, Bank Cottage, Great Comberton, Worcestershire WR10 3DP

Northern Horticultural Society
Harlow Carr Botanical Gardens, Crag Lane, Harrogate, N Yorks HG3 1QB

RHS Lily Group
Dr A. F. Hayward, Rosemary Cottage, Lowbands, Red Marsley, Glos. GL19 3NG

Scottish Rock Garden Club
General Secretary: Dr Evelyn Stevens, The Linns, Sheriff Muir, Dunblane, Perth FK15 0LP

SUPPLIERS

Many of the popular bulb catalogues carry advertisements for several of the commoner and easy species. Growing these species from seed would be time-consuming and probably more expensive than buying the flowering size bulbs.

Specialist bulb suppliers are slowly adding more species of *Allium* to their lists. Those marked with an asterisk in the list below supply bulbs to customers outside the UK. Some of the general catalogues also carry a limited number of *Allium* for sale. Unfortunately, many smaller nurseries no longer sell by mail order.

The Plant Finder, compiled by Chris Philip and Tony Lord, published by Headmain Ltd for the Hardy Plant Society, is a guide to over 40,000 plants including *Allium* (135 entries in the 1991/2 edition).

Jacques Amand Ltd,*
The Nurseries,
145 Clamp Hill,
Stanmore, Middlesex HA7 3JS

Avon Bulbs,*
Burnt House Farm,
Mid Lambrook,
South Petherton,
Somerset TA13 5HE

Walter Blom & Son Ltd,*
Coombelands Nurseries, Leavesden,
Watford, Herts. WD2 7BH

Blooms of Bressingham,*
Diss, Norfolk IP22 2AB   Export

Broadleigh Gardens,*
Barr House, Bishops Hull, Taunton,
Somerset TA4 1AE   Export

Rupert Bowlby,
Gatton,
Reigate, Surrey RH2 0TA

Cambridge Bulbs,*
40 Whittlesford Road,
Newton, Cambs. CB2 5PH   Export

P.J. & J.W. Christian,*
PO Box 468,
Wrexham, Clwyd, Wales LL13 9XR Export

Peter J. Foley,*
Holden Clough Nursery, Bolton-by-
Bowland,
Clitheroe, Lancs. BB7 4PF   Export

Monocot Seeds (and plants) (M.R. Salmon)*
Jacklands, Jacklands Bridge, Tickenham,
Clevedon, Avon BS21 6SG   Export

Paradise Centre,
Twinstead Road, Lamarsh,
Bures, Suffolk CO8 5EX

Potterton & Martin,*
The Cottage Nursery,
Moortown Road, Nettleton,
Nr Caistor, North Lincs. LN7 6HX

Van Tubergen UK Ltd,*
Bressingham, Diss,
Norfolk IP22 2AB

Wallace & Barr Ltd,*
The Nurseries,
Marden, Kent TN12 9BP

Advertisements for nurseries appear in specialist journals. *Gardening By Mail* is a guide for gardeners in the USA.

APPENDIX IV *Hardiness Zone Maps*

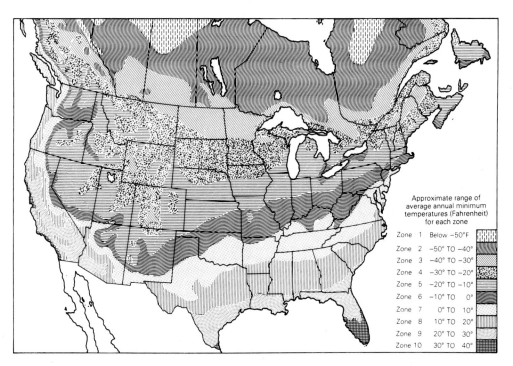

Approximate range of
average annual minimum
temperatures (Fahrenheit)
for each zone

Zone   1   Below −50°F

Zone   2   −50° TO −40°

Zone   3   −40° TO −30°

Zone   4   −30° TO −20°

Zone   5   −20° TO −10°

Zone   6   −10° TO   0°

Zone   7   0° TO   10°

Zone   8   10° TO   20°

Zone   9   20° TO   30°

Zone 10   30° TO   40°

Hardiness zones of North America

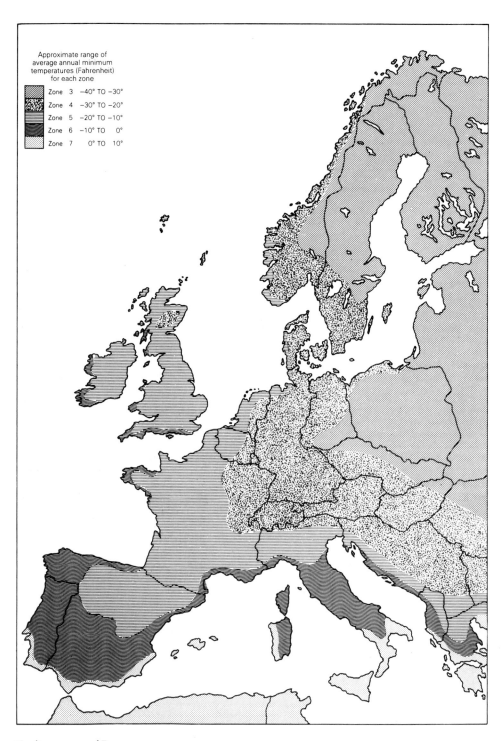

Approximate range of
average annual minimum
temperatures (Fahrenheit)
for each zone

Zone 3   −40° TO −30°
Zone 4   −30° TO −20°
Zone 5   −20° TO −10°
Zone 6   −10° TO   0°
Zone 7   0° TO 10°

Hardiness zones of Europe

Select Bibliography

Allium literature is mentioned in the text. While not attempting to compile a complete list of references, some of the following have provided an enjoyable respite, disguised as research.

Anderson, Berta (1976) *Wild Flower Name Tales*

Barber, R. (ed. trans.) (1979) *Life & Campaigns of the Black Prince*

Barrow, John (ed.) (1860) *Captain Cook's Voyages of Discovery*

Bordia, A., Bansal, H.C., Arora, S.K. & Singh, S.V. (1975) *Effect of the Essential Oils of Garlic and Onion on Alimentary Hyperlipemia*

Bothwell, J. (1950) *The Onion Cookbook*

de Candolle, A. (1855) *Géographie Botanique Raisonnée*

Chief, Roberto (1984) *MacDonald Encyclopedia of Medicinal Plants*

Clapham, Tutin and Moore *The Flora of the British Isles* (CUP 1989)

Clarkson, R.E. (1939) *Magic Gardens*

Dayton, W.A. (1960) *Notes on Western Range Forbs, Agriculture* (US Department of Agriculture, Handbook 161)

Dalechamp, J. (1587) *Historia Generalis Plantarum*

Devoto, Bernard (ed.) (1953) *The Journals of Lewis and Clark*

Diaz, Bernal (1963) *The Conquest of New Spain*, ed. trans. Cohen, J.M.

Dioscorides Pedanius (c.AD70) *The Greek Herbal*, ed. Gunther, R.T. (1934)

Fitzgibbon, T. (1972) *A Taste of the West Country*

Fulder, S. (1989) *Garlic and the Prevention of Cardiovascular Disease*

Gerard, J. (1597) *Herball or Generall Historie of Plantes*

Grant, Michael (1954) *Roman Literature*

Hadfield, Miles (1955) *Pioneers of Gardening*

Herodotus *The Histories*, trans. (1954) De Selincourt, A.

Howey, P. *Roman Cook Book*

Hutchinson, J. (1973) *The Families of Flowering Plants*

Jones-Mortimer, H.M.C. (1948) *Wales and the Welsh Guards*

Keen, B. (ed./trans.) (1959) *Life of the Admiral Christopher Columbus by his Son Ferdinand* (c.1530)

Kendler, B.S. (1987) *Garlic (Allium sativum) & Onion (A. cepa): A Review of their Relationship to Cardiovascular Disease*

Kimball, Y. & Anderson, J. (1965) *The Art of American Indian Cooking*

Kirk, D.R. (1975) *Wild Edible Plants of the Western United States*

Li Shih-Chen (c.1570) *Chinese Medicinal Herbs*, trans. Porter Smith, F. & Stuart, G.A.

McMahon, B. (1806) *American Gardeners' Calendar*

Mann, L.K. & M. (1960) *Decorative Onions*

Michaud, J. (1853) *History of the Crusades*

Pliny the Elder *Natural History*, trans. (1951–1963) Jones, W.H.S., Vols. 6, 7 and 8

Romans, B. (1775) *Concise Natural History of East & West Florida*

Soyer, A. (1853) *The Pantropheon*

Stearn, W.T. (1966; 3rd ed. rev. 1983) *Botanical Latin*

Sturtevant, E.L. (1972) *Edible Plants of the World*, ed. Hedrick, U.P.

Turner, N.J., Thomas, J., Carlson, B.F. & Ogilvie, R.T. (1984) *Ethnobotany of Nitinaht Indians of Vancouver Island*

Turner, N.J., Bouchard, R. & Kennedy, D.I.D. (1981) *Ethnobotany of Okanagan-Colville Indians of British Columbia and Washington*

Vischer, I. (1953) *Now to the Banquet*

Glossary

ACUMINATE drawn out at the apex into a gradually tapering point.

ACUTE terminating in a sharp or well-defined point.

ADNATE fusion of dissimilar parts or adhering along length to dissimilar part.

ALTERNATE leaves arranged alternately along the stem, not whorled or opposite.

ANNULUS a ring of tissue.

ANTERIOR in flowers the side facing away from the axis of the inflorescence.

ANTHER the pollen-bearing organ of the flower.

ANTHESIS flowering time.

APEX, APICES tip of stem or leaf.

APICULATE with a small, broad point at the apex.

APPENDAGE extra attachment to an organ, often apparently useless.

ARIL A fleshy, usually coloured, covering or appendage on a seed.

ATTENUATE drawn out into a long slender tip.

AXIL the junction between leaf and stem.

BASAL leaves arising at ground level or below, not from an aerial stalk.

BEAK a pointed extension to bracts, seeds, ovaries or capsules.

BEARDED having a tuft of long hairs.

BIDENTATE 2-toothed.

BIFID cleft in two.

BINOMIAL a name having two parts.

BLADE the expanded part of the leaf = lamina: excludes the stalk.

BLOOM whitish waxy or powdery (glaucous) covering on leaf or stalk.

BRACT a much reduced leaf; usually subtending a flower.

BRACTEATE having bracts.

BRACTEOLE extra, smaller bract, often enclosed within the main bract.

BULB underground bud or stem with fleshy coats or scales.

BULBIL small bulb produced in the flowering head or in leaf axils.

BULBLET small bulb produced alongside the parent bulb.

CAESPITOSE growing in clumps.

CALYX outer set of perianth segments; collective name for the sepals.

CAMPANULATE bell-shaped.

CANESCENT having a greyish, hoary grouping of short hairs.

CAPSULE dry fruit of at least 2 carpels, splitting open when ripe.

CARPEL one female, seed-bearing unit of a flower.

CAULINE borne on the stem above ground; opposite of basal.

CERNUOUS nodding.

CILIATE marginally fringed with hairs.

CLAVATE club-shaped.

CLEFT deeply cut.

COHERENT having similar parts united.

COMPRESSED flattened, usually laterally.

CONNATE united by fusion of similar parts, e.g. 2 anthers joined at the bases.

CONNIVENT coming into contact, converging; used of similar organs.

CORDATE heart-shaped.

CORIACEOUS leathery in texture.

CORM enlarged fleshy base of a stem, bulb-like but solid.

CRENATE with rounded, marginal teeth.

CRENULATE finely crenate.

CREST elevated, toothed or irregular ridge of tissue.

CULTIGEN plant/group, of apparent specific rank, known only in cultivation.

CUNEATE wedge-shaped.

CUSPIDATE spear-shaped at the tip.

DEFLEXED bent sharply downwards.

DEHISCENT splitting open at maturity.

DENTATE toothed, with the teeth directed outwards.

DENTICULATE minutely dentate.

DEPRESSED flattened vertically or at the apex.

DISJUNCT occurring in two widely separated geographic areas.

DISTINCT separate, not united.

DORSAL back or upper surface, i.e. upper surface of a leaf.

DORSIFLEXED used of an anther attached by its back, not by its lower edge, to the filament.

E- OR EX- without, lacking or outside.

EMARGINATE having a shallow notch at the tip.

ENDEMIC found as native in one country or area only.

ENTIRE having an even margin, not notched or toothed.

EROSE having the margins irregularly denticulated as if bitten by an animal.

EXSERTED projecting beyond the enveloping organs.

FALCATE sickle-shaped.

FAMILY a group of related genera.

FARINOSE with a mealy or powdery covering.

FIBRILLOSE shredding into fine fibres.

FIBROUS (1) slender, non-swollen roots; (2) mat of fibres in a bulb coat.

FILAMENT the stalk supporting an anther.

FILIFORM threadlike.

FIMBRIATE fringed.

FISTULOUS in the form of a hollow tube closed at both ends.

FLEXUOUS wavy, curved, zigzag.

FLORET a small flower (normally applied to flowers of grasses or composites).

FREE not joined together, often used when referring to tepals.

GAMOPETALOUS corolla in one piece, petals united at least at base.

GENUS (1) smallest natural group of related but distinct species; (2) group of plants all bearing the same first name; second part of binomial indicates the species.

GIBBOUS swollen on one side.

GLABRATE, GLABROUS almost hairless or smooth.

GLAND an organ of secretion (occasionally used for flower parts of unknown function).

GLANDULAR having glands, sticky.

GLAUCESCENT verging upon or becoming glaucous.

GLAUCOUS having a bloom or whitish covering.

GLOBOSE spherical.

GLOMERATE crowded into a compact, spherical cluster.

GYNOECIUM collective name for all the carpels in a single flower.

HERBARIUM a collection of dried specimens.

HIRSUTE clothed with coarse, stiff hairs.

HISPID clothed with bristle-like hairs.

HYALINE transparent or translucent.

HYBRID cross between two different plant taxa; may be natural or artificial.

HYPOGYNOUS flowers with stamens, petals and sepals attached below the ovary.

IMBRICATE overlapping, like roof tiles.

INCISED cut deeply, sharply and irregularly into lobes or segments.

INCLUDED not protruding beyond, e.g. stamens concealed within the perianth.

INDURATE hardened.

INFERIOR referring to the ovary sited below the rest of the flower parts.

INFLORESCENCE flower cluster.

INTRORSE facing inwards, e.g. anthers shedding pollen from inward surface.

INVOLUTE having the edges rolled inwards.

KEEL a prominent rib, ridge or crease on the undersurface.

LANCEOLATE long and narrow but broadest at the base.

LENTICULAR lens-shaped, with two convex sides.

LINEAR long and narrow with parallel margins.

LINEAR-LANCEOLATE slender, tapering to a point at the tip.

LOCULE, LOCULUS one of the cavities or chambers in an ovary or anther.

MEMBRANOUS, MEMBRANEOUS dry, thin and semi-transparent.

MONOGRAPH written account of a single genus or family.

MUCRONATE having a minute and abrupt point at the apex.

NECTARY nectar-producing organ, usually near the tepal base.

NERVE prominent vein.

NETTED net-like arrangement, usually referring to the bulb coat fibres.

OB- prefix meaning inversely or with narrower end towards attachment point.

OBLANCEOLATE lance-shaped but wider above the middle.

OBLONG rectangular but with rounded corners.

OBOVATE egg-shaped in reverse, wider above the middle; in the flat.

OBOVOID three-dimensional obovate.

OBTUSE blunt or rounded at the end.

OFFSET bulblet produced vegetatively at the base of parent bulb.

OPPOSITE originating in pairs, e.g. leaves on a stem.

ORBICULAR round, circular.

OVARY part of the carpel or gynoecium which contains the seeds.

OVATE egg-shaped and broadest at the base; in the flat.

OVOID egg-shaped, relating to solid objects.

OVULE the seed before fertilisation.

PAPILLAE small, elongated protuberances.

PAPILLOSE minutely warty or pimply.

PEDICEL the stalk of a single flower.

PEDICELLATE having a flower stalk.

PERIANTH collective name for the petals and sepals.

PERICARP the ovary wall.

PERSISTENT lasting, not deciduous, remaining attached to the stem.

PETAL one of the white or coloured, inner perianth segments.

PETALOID resembling a petal.

PETIOLATE having a leaf stalk.

PETIOLE leaf stalk.

PISTIL female organ of flower, when complete including ovary, style and stigma.

POSTERIOR side of flower facing the axis of the inflorescence.

PUBERULENT very minutely hairy.

PUBESCENT hairy.

PYRIFORM pear-shaped.

RECEPTACLE often thickened, upper part of stem carrying the flowers.

RECURVED moderately bent backward in a curve.

REFLEXED bent abruptly downward.

RETICULATE marked with a network of veins.

REVOLUTE margins rolled back or under, opposite of involute.

RHIZOME prostrate underground stem or branch; unlike a root has nodes.

RIB prominent ridge or vein.

RUDERAL growing in rubbish or in waste places.

RUGOSE wrinkled.

SACCATE bag-shaped.

SCABROUS rough to the touch.

SCALE (1) fleshy individual segment of a bulb; (2) reduced leaf.

SCAPE leafless flower stalk, rising from ground or cluster of basal leaves.

SCAPOSE bearing a scape.

SCARIOUS thin, dry, scale-like, not green.

SECUND 1-sided, turned to one side.

SEED ripened ovule.

SEPAL an outer perianth segment, a segment of the calyx.

SEPTATE divided into partitions.

SERRATE having sharp teeth pointing forward, saw-toothed.

SERRULATE minutely serrate.

SESSILE lacking a stalk.

SHEATH basal portion of leaf, the part surrounding the stem.

SINUATE wavy-margined.

SINUS cleft or indentation between lobes.

SPATHE sheath enclosing the flower buds.

SPECIES classification of groups, distinct, usually capable of interbreeding.

SPP. species, plural.

SSP. subspecies.

STAMEN pollen-producing organ of a flower.

STELLATE star-shaped.

STERILE flower incapable of bearing seeds.

STIGMA apex of style, which receives the pollen.

STRIATE marked with fine, longitudinal furrows or lines.

STRICT upright, narrow, very straight.

STYLE the slender, upper part of a carpel.

SUB- prefix meaning almost.

SUBGLOBOSE approaching spherical.

SUBSP. subspecies.

SUBSPECIES subdivision of a species.

SUBTEND to occur below, e.g. a bract subtends a flower.

SUBULATE awl-shaped, tapering from base to apex.

SUPERIOR referring to ovaries sited above the base of other flower parts.

SYNONYM a superseded name, no longer correctly in use.

TAXON (pl. taxa) any named taxonomic entity, e.g. genus, family or species.

TAXONOMY classification.

TEETH any tooth-like organ; may refer to jagged margins.

TEPAL the floral leaves of a monocotyledon flower = perianth segments = dicotyledon sepals and petals.

TEPAL-LOBE free, unfused parts of a tepal.

TERETE cylindrical, circular on cross section, without ridges or grooves.

TERNATE divisions or groups of three.

TRIFID 3-clefted but not to the base.

TRIGONOUS 3-angled, 3-sided, solid body with plane faces.

TRIQUETROUS 3-cornered.

TUBER swollen, scale-less underground organ of solid tissue.

TUBERCLE small ovoid or spherical swelling.

TUNIC covering coat of bulb or corm.

UMBEL inflorescence, usually flat-topped, in which the flower stalks all spring from roughly the same point in an axis.

URCEOLATE urn-shaped; wider in middle than above or below.

VALVE one of the pieces into which a capsule or bract splits.

VAR. variety.

VARIETY lower rank than species/subspecies.

VENTRAL on the side facing the axis, undersurface.

VIVIPAROUS producing young plants in place of flowers.

WIDESPREAD distributed over a wide area but not necessarily common locally.

XEROPHYTE plant adapted to very dry conditions.

ZYGOMORPHIC bilaterally symmetrical.

A. polyastrum (Author)

General Index

Common names and other genera mentioned in the text are listed, attributions will be found in species' descriptions. Dates are given wherever possible, being omitted for the living. Localities are also unlisted unless of specific interest.

Aase, Hannah 63
Abrams, LeRoy
  1874–1956 43, 110
Agrippa, Marcus Vipsanius
  63–12BC 21
Airy Shaw, H. K.
  1902–1985 40, 106
Aitchison, Surgeon Major
  James, c.1832–1896 38
Albury, Sydney d.1970
  115
Alexander the Great
  356–323BC 20, 21
Alpine Garden Society &
  Bulletins 10, 38, 43, 85,
  110, 115, 116, 117, 152
Amaryllis caspia 29, 80
American Rock Garden
  Society & Bulletins 10,
  43, 44, 110, 115, 152
Anderson, E. Bertram,
  1885–1971 41, 76, 147
Anglo-Saxons 14, 15
aphids 50–2
Apicius, AD14–37 14, 20,
  26
Archibald, J. C. A. 38, 116
Askalon, 69
Askalonium krommoon
  20
Aspen Onion 72
Augustus 63BC–AD14 26,
  68

Babington, Charles
  Cardale, 1808–1895 70
Bailey Hortorium 43
Bailey, Liberty Hyde
  1858–1954 32
Baileya 23, 44
Baker, John Gilbert
  1834–1920 42, 93
Balls, Edward Kent c.
  1890–1984 38
Balzac, Honoré de
  1899–1850 27
Bauer, Franz 1758–1840
  40
Baxter, Felicity 38, 119
Bear's Garlic 25
Beckett, Kenneth A. 38
Bees Nursery 71
Berkeley University Botanic
  Garden 101, 152
Black Prince 1330–1376
  22
Blamey, Marjorie 40
Blanchard, Keith 41

Boissier, Pierre Edmond
  1810–1885 20, 40, 84
Bolander's Onion 72
Botrytis 51
Bowles, E. Augustus
  1865–1954 53, 57, 68
Breton onion sellers 14
Brewer's Onion 74
Brickell, C. D. 38
Brunfels, Otto 1489–1534
  39
Buckrams 25
Buczaski, S. 52
Bulley, Arthurt Kilpin
  1861–1942 71

Calendar of the Hsia 15
Caloscordum nerinifolium
  112
Camassia leichtlinii 15
Camassia quamash,
  quamash 15
Campbell, E. 40, 41
Canada Garlic 78
Candide 16
Candolle, Alphonse de
  (A.D.C.) 1806–1893
  20, 21
Candolle, Augustin P. de
  (D.C.) 1778–1841 41, 80
Canterbury Tales 26
cardiovascular disease 17,
  19
Catawissa Onion 22
Charlemagne 742–814 20,
  23
Chaucer, Geoffrey,
  c.1343–1400 26
Cheese, Martin 115
Chicago 23, 78
Chinese Chives 5
Chittenden, Frederick
  James, 1873–1950 41
chives 18, 20, 22, 23, 27,
  53
Choctaw Indians 16, 22
Christian, Paul 38, 115
Clements, Julia 84
Clifford, Harold Trevor
  29
Clusius, Carolus (Charles
  de L'Ecluse) 1526–1609
  39, 86, 128
Coastal Onion 91
Columbia River Gorge
  125
Columbus, Christopher
  1451–1506 21

Columella, L. Junius
  Moderatus c.AD40 21
Cook, Capt. James
  1728–1779 16
Cornish tin miners 27
Cortés, Hernando
  1485–1547 21, 22, 23
Cowan, James c.1830 87,
  97
Crecy, Battle of 1346 22
Crinkled Onion 87
Crinum caspium 80
Cronquist, Arthur 43
Crow Garlic 25, 39, 145
Cunnington, Peter 38
Cuthbert's Onion 88
Cybele 13

Dadd, R. 98
daffodils, emblem 15
Dahlgren, Rolf Martin
  Teodor 29
Dalechamp, Jacques
  1513–1588 21
Dana, Richard Henry Jr
  1815–1882 93
Davies, Pat 98, 152
Davis, Peter 38, 40
Death Camas 15
Death Valley 70
Delia antiqua 51
Delile, Alire Raffeneau
  1778–1850 41
diallyl disulphide, diallyl
  trisulphide 19
Diels, Friedrich Ludwig
  Emil, 1874–1945 106
Dioscorides, Pedanios, first
  century AD 21, 22
Dodoens, Rembert
  (Dodonaeus)
  c.1518–1585 22
Don, George the Younger
  1798–1856 36, 39, 64
Douglas, David
  1798–1834 38, 92, 94
Douglas's Onion 92
Downy Mildew 52

Edward III 1312–1377 22
Éelworms 51
Egypt 12–14, 21, 22, 50
Egyptian/Tree Onion 20,
  21, 22
Ekberg, Lars 42
Elliott, Dr J. G. 77
Elwes, Henry John
  1846–1922 106

European Garden Flora
  64, 114, 137
Ever Ready Onion 22

Farrer, Reginald
  1880–1920 37, 42, 89,
  90, 110, 141
Fedtschenko, Aleksei P.
  1844–1873 38, 39
Fedtschenko, Boris
  1873–1947 39
Fedtschenko, Olga
  1845–1921 38, 39
Feinbrun, Naomi 35
Field Garlic 24, 25, 114
Flintoff, Jerry 115
Flora Europaea 8, 30, 31,
  34, 40, 80, 96, 117, 136
Flora Iranica 8, 32, 36,
  38, 40, 74, 77, 121, 129
*Flora Reipublicae Popularis
  Sinicae* 35, 42, 97, 120,
  139, 145, 147
Flora Turkey 8, 40, 46,
  85, 86, 118
Flora USSR 8, 36, 40, 64,
  65, 71, 72, 74, 75, 80,
  112, 121, 126, 129, 136,
  137
Fluellen 22
Forrest, George
  1873–1932 37, 38, 42,
  66, 71, 97, 106, 120,
  147
Fortune, Robert
  1812–1880 37
Franklin, Benjamin
  1706–1790 18
Fringed Onion 95
Fuchs, Leonhard
  1501–1566 39
Furse, Mrs Polly 38
Furse, Rear Admiral Paul
  1904–1979 38, 74, 77,
  107, 119, 124

Galen c.130–c.200 17
garlic 12–19, 22–3, 27,
  28, 30, 36, 37, 39, 57,
  78, 127–8, 157
Gerard, John 1545–1607
  22, 23, 25, 26, 27
Geyer, Carl Andreas
  1809–1853 98
Geyer's Onion 98
Gooding, Leslie M. 100
Great Headed Garlic 23,
  67

Grey, Charles Harvey (born Hoare) 1875–1955 40, 69, 78, 79, 80, 82, 84, 89, 106, 110, 113, 136, 137, 140, 146
Grey-Wilson, Christopher 38, 40, 41, 85, 134
Gypsy Onion 25

Hadrian AD76–138 14
Haller, Albrecht von 1708–1777 39
Halliwell, Brian 38
Han Period dictionary c.206BC–AD220 142
Hanelt, Peter 147, 152
Hara, Hiroshi 1911–1986 42
Hardy Plant Society 2, 8, 153
Harkness, Bernard 1907–1988 8
Hedge, Ian Charles 74
Henry IV 1367–1413 26
Henry V 1387–1422 22, 27
Herod the Great 74–4BC 69
Herodotus c.480–c.425BC 12
Hill, Jason (Dr F. A. Hampton)c.1930 86
Hippocrates c.460–c.377BC 21
Hitchcock, C. Leo 1865–1935 43
Hog's Garlic 25
Holmgren, A. H. 43
Holmgren, N. J. 43
Holmgren, P. K. 43
Homer c. tenth century BC 13
honey 17, 23, 27
Hoog, Michael 108
Hooker, Sir Joseph Dalton, 1817–1911 42
Hooker, Sir William Jackson, 1785–1865 98
Hooker's Onion 63
Horace 65–8BC 12
Horton, S. Victor, 1919–1980 38, 115, 116, 117
Humboldt, Baron Alexander von 1769–1859 21
Hurricane Ridge, Olympic Mts, Washington State 87
Hutchinson, John 1884–1972 29

Janka, Victor de 1837–1890 41
Japanese Bunching Onion 24, 59, 95–6
John of Gaunt 1340–1399 26

Jones's Onion 94
Juvenal c.AD60–140 26
Keble Martin, William 1877–1969 35, 114
Keeled Garlic 60, 79
Kingdon-Ward, Frank 1885–1958 78, 79, 84
Kirghis 80
Kiu ts'ai 142
Kollmann, Fania 40, 41, 86, 97
Komarov, Nikolai Fedrovi 1901–1942 40
Koyuncu, M. 41, 97
Lancaster, Roy 38
leek 12, 14–16, 19–23, 26, 27, 39, 57, 59, 62, 67, 121
Levant Garlic 23
Léveillé, Hector 1796–1870 106
Lewis & Clark Expedition 1804–6 15
Li Shih-Chen, 1518–1593 19
Lily Beetle 51
Lily Group, R. H. S. 35, 41, 42, 74, 86, 153
Linnaeus (Linne, Carl von) 1707–1778 14, 39, 82, 124
liquamen 26
Liverpool University Botanic Garden, Ness 71
Lotus Eaters 13
Loudon, John Claudius 1783–1843 21
Macdonald Encyclopedia 17, 18, 157
Maclean, Colonel c.1880 106
Macoun, John 1832–1920 145
MacPhail, Jim 38
Maire, R. C. J. Edouard 1878–1949 38, 106
manna 14
Marquette, Père Jacques 1637–1675 23
Martial, (Marcus Valerius Martialis) AD c.42–102 20, 26
Maslin, Paul 1909–1984 64
Mathew, Brian 38, 40, 41, 124, 133
Mattioli, Pietro Andrea 1500–1577 39
Maximowicz, Carl (K.J.) 1827–1891 38, 42
MacMahon, Bernard 1775–1813 20
Meadow Leek 78
Medusa 78

Melanocrommyn section 25
Menzies, Archibald 1754–1842 63
mice 50
Michaud, Joseph 1767–1839 20, 157
Milula 42, 120
moles 50
Moore, Harold Emery 1917–1980 23, 44, 109
Morganwg, Iolo (Edward Williams) 1746–1826 15
Morton, Conrad Vernon 35, 110
Mountain Garlic 133
Mouse Garlic 23
Multiplier Onion 22
Munz, Philip Alexander 43

name derivations 15
Naples Garlic 111
narcissus 110, 111
Narrow Leaved Onion 68
National Collection of Alliums 152
Nectaroscordum siculum ssp. *bulgaricum* 74
Nectaroscordum siculum ssp. *siculum* 134
Nero AD37–68 22
Nez Perce Indians 15
Niehaus, T. F. 44
Nitinaht Indians 16, 157
Nodding Onion 53, 81
Northern Horticultural Society 153
Nuttall, Thomas 1786–1859 38, 93

Odysseus 13
Ohwi, Jisaburo 1905–1977 35, 43
Okanagen-Colville Indians 16
One Leaved Onion 142
onion 12–16, 18–19, 20–1, 26, 27, 28, 30, 32, 39, 50, 51, 80–1
Onion Fly 51
Onion Mildew 52
Ovid, (Publius Ovideus Naso) 43BC–AD17 20
Ownbey, Francis Marion 1910–1974 43, 63, 82, 99, 100, 109, 110
Özhatay, N. 41

Pace eggs 19
Pallas Athene 78
Paper Onion 68
Parkinson, John 1567–1650 39, 108
Peninsular Onion 120
Peronospora destructor 52
Perseus 78

Phillips, Roger 41, 115
Pike, Zebulon M. 1779–1813 98
plague 17, 52
Pliny the Elder AD23–79 20, 21, 22, 26, 157
Pliny the Younger AD62–113 13
Poitiers, Battle of, 1356 22
Polo, Marco c.1254–1324 12
Polunin, Oleg 1914–1985 38, 41, 43
Potanin, Grigori Nicolaevich 1835–1920 89
Potato Onion 20, 22
Prairie Onion 93, 139
Preston, Lancashire 19
Przewalski, Nikolai Mikhailovich 1839–1888 38, 42, 84, 89
Purdom William 1880–1921 37, 42, 89, 90
Purdy, Carlton 1861–1945 38
pyramids 12

quamash, *Camassia quamash* 15
Queen of Sheba 14

rakkyo 82, 135
Ramps 25, 59, 140
Ramsoms 18, 39, 59, 140, 143
Rechinger, Karl Heinz 40
Redouté, Pierre Joseph 1761–1840 40
Regel, Albert von 1845–1908 38
Regel, Eduard A. von 1815–1892 8, 38, 39, 42, 74, 93, 109, 119, 124, 126, 137
Reichenbach, H. G. Ludwig 1793–1879 41
Retzius, Anders Johan 1742–1821 39
Reveal, J. L. 43
Richard II 1367–1400 26
Rickett, Harold William 1896–1989 43, 109
Ripper, C. L. 44
Rix, Martyn 38, 41, 45
Rocambole 19, 131
Roche, Francois de la (d.1813) 41
Rock, Joseph 1884–1962 84
Rock Onion 106
Roderick, Wayne 38
Romans, Bernard 1720?–1784? 16, 157
Rose Leek 78
Rosy Flowered Garlic 24

Round Headed Garlic 25
Round Headed Leek 135
Royal Botanic Garden,
  Edinburgh 39, 41, 55,
  79, 107, 152
Royal Botanic Garden,
  Kew 7, 29, 42, 76, 86,
  91, 98, 99, 107, 121,
  123, 128, 146, 152
Royal Horticultural Society
  & Journals 38, 41, 51,
  52, 77
Rust fungus 29, 52, 87
Rydberg, P. Axel
  1860–1931 98

Sand Leek 19, 131
Satan 21
Savile, D. B. O. 29
Scallions 20
Schenk, George 140
Scilla paradoxus 119
Sclerotium cepivorum 51
Scottish Rock Garden
  Club 10, 153
Scythe Leaved Onion
  94
Seneca, Lucius Annaeus
  c.BC–AD65 26
Serpent Garlic 86
Shakespeare 1564–1616
  22
Shallots 20, 69
Shortstyle Onion 73

Sibthorp, John
  1758–1796 40
Sierra Onion 77
Sirak -e chak piaz 128
slugs 47, 49, 55, 66, 90,
  107, 135
Small, John Kunkel
  1869–1938 141
Small Yellow Onion 96
Smith, Sir James Edward
  1759–1828 40
Smut 51
Species Plantarum 39
St David 21
Stag's Garlic 25
Stainton, J. D. A. 43
Stearn, William T.8, 23,
  40, 41, 42, 64, 86, 90,
  106, 119, 120, 123,
  157
Stevenson, Robert Louis
  1850–1895 19
Stoker, Bram 1847–1912
  17
Striped Garlic 88
Sullivan, Brig. General
  John 1740–1795 21
Swynford, Katherine
  1350–1403 26
Synge, Patrick 1907–1982
  38, 110

Tall Swamp Onion 144
Tapertip Onion 63

Tennyson, Alfred, Lord
  1809–1892 13
Tenochtitlán 21
Theophrastus
  c.370–285BC 20, 21
Thompson, John William
  1890–? 43
Threave Gardens 72,
  152
thrips 51
Thunberg, Carl Pehr
  1743–1828 124
Tiberius 42BC–AD37 26
Top Onion 22
toxicity 19, 49
Trajan AD52–117 14
Tree Onion 21
Tutin, Thomas Gaskell
  1908–1987 40
Two Years before the Mast,
  1840 93
Urocystis cepulae 51
Valak 91, 101, 113
Validallium tricoccum 141
vampires, ghouls 16–17,
  145
Van Tubergen Nursery 38,
  64, 123, 154
Vancouver, Captain
  George 1758–98 63
Villar (Villars), Dominique
  1745–1814 110
virsues 50, 52

Vlad the Impaler, fifteenth
  century 17
Vvedensky, Aleksei
  Ivanovich 1898–1972
  35, 40, 42, 74, 121, 129

Wallich, Dr Nathaniel
  1786–1854 146
Walters, S. M. 40
Watson, John 39, 115
Weevils, Clay-Coloured,
  Vine 51, 52
Welsh emblems &
  derivations 14, 15, 22
Welsh Onion 15, 24, 59,
  66, 95–7
Welsh Regiment 15, 157
Wendelbo, Per Erland Berg
  1927–81 38, 40, 42, 43,
  72, 74, 93, 117, 121, 129
White Rot 52
Wild Garlic 16, 78
Wild Leek 23, 67, 140
Wild Onion 23
Williams, L. H. J. 42
Wilson, Edward
  1876–1930 37, 42, 84
witchcraft 16
Wood Garlic 143

xonocatl 21
Xu, J. M. 42, 147
Yeo, Peter 29

Ziqadenus venenosus 15

Index of Species

Synonyms have not been differentiated, as all authorities do not agree on species' status. Plates and line drawings are in bold type

Allium

aaseae 63
achaium 109
acidoides 107, 147
acuminatum 23, 37, 55,
  62–3, 64, 109, 110
acutiflorum 64, 150
affine 146
aflatunense 36, 53,
  57, 64–5, 123, 137,
  152
aitchinsonii 38
akaka 23, 36, 38, 40, 56,
  65, 72, 91
albidum 65, 148
  subsp. albidum 65
  subsp. caucasicum 65
albiflorum 150
albopilosum 66, 83, 84

album 112
alleghemiense 82
altaicum 59, 66, 96
altissimum 66, 104
amabile 38, 55, 56, 58,
  66–7, 67, 106–7, 147
ambiguum 126
amblyanthum 117
amethystinum 67, 150
ammophilum 65, 66
amoenum 126
ampeloprasum 14, 22, 23,
  35, 36, 67–8, 68, 69,
  121, 150
  var. atroviolaceum 70
  var. babingtonii 57, 59,
    62, 67, 70
  var. bulbiferum 60, 68

amplectens 68
andersonii 134
angulosum 23, 24, 59,
  68–9, 123, 133, 137,
  148
  var. caucasicum 66
  var. minus 134
angustitepalum 126
angustoprasum 23
anisopodium 69
arenicola 98
aridum 139
arvense 135
  var. trachypus 135
ascalonicum 20, 69
aschersonii 37
assimile 146
atriphoeniceum 79

atropurpureum 69, 151
  var. hirtulum 102
  var. inyonis 70
atrorubens 70
  var. inyonis 70
atroviolaceum 70, 150
attenuifolium 68
auctum 112
austinae 78
azureum 57, 75

babingtonii 35, 68, 70
baeticum 150
baicalense 134
baissunense 80
bakeri 83
barszczewskii 9, 70–1,
  124

beckerianum 103
beesianum 36, 37, 38, 49, 55, 56, 58, 60, 71, 71–2, 89, 97, 135, 152
bidwelliae 72, 78
bisceptrum 72
blandum 80
bodeanum 72
bolanderi 72–3
  var. *stenanthum* 73
bornmuelleri 109, 150
bourgeaui 68, 150
brahuicum 80
brandegei 95
breviradium 149
brevistylum 23, 37, 58, 73, 73–4, 100, 145
breweri 74, 94
bucharicum 55, 56, 65, 74, 129
  PF 6268 74
buhseanum 129
bulgaricum 74–5
bullardii 72
bulleyanum 147

caeruleum 36, 57, 60, 75, 75–6, 146, 150
  var. *bulbilliferum* 60, 146
caesium 36, 57, 60, 62, 75
callimischon 36, 56, 56, 76, 77, 88, 150
  subsp. *callimischon* 76
  subsp. *haemostictum* 62, 76
callistemon 97
calocephalum 60, 76–7, 77
campanulatum 37, 72, 77, 77–8
canadense 15, 21, 23, 37, 60, 78, 93
  var. *canadense* 78
  var. *fraseri* 78
  var. *lavendulare* 78
  var. *mobilense* 78
  var. *ovoideum* 78
  var. *robustum* 78
candolleanum 147
capillare 109
caput medusae 78–9, 79
cardiostemon 56, 79
carinatum 35, 47, 59, 59, 60, 79, 97, 114, 131, 146, 150
  subsp. *carinatum* 79
  subsp. *pulchellum* 53, 58, 59, 79, 97, 115, 118, 122–3
  subsp. *pulchellum* 'Album' 122
  subsp. *pulchellum* Van Tubergen's variety 123
carneum 126

carolinianum 41, 80
cascadense 87
caspicum 80
caspium 80, 151
caucasicum 128
cepa 20, 21, 69, 80–1, 97, 117, 148
  ascalonia 20
  var. *aggregatum* 22
  var. *bulbiferum* 22
  var. *multiplicans* 22
  var. *proliferum* 22
  var. *solaninum* 22
  var. *sylvestre* 20, 117
  'Viviparum' 21
cernuum 15, 16, 23, 24, 37, 53, 59, 60, 81–2, 136, 152
  var. *neomexicanum* 82
  var. *obtusum* 82
chamaemoly 56, 82, 149
chamaespathum 150
charaulicum 127
chinense 24, 82–3
chionanthum 147
christophii 36, 38, 55, 58, 62, 66, 72, 83, 83–4, 99, 152
chrysantherum 60, 84
chrysanthum 60, 84
chrysanthemum 60, 85, 109, 150
ciliatum 138
cilicicum 85
circinatum 85
circinnatum 85, 149
clusianum 39, 138
coeruleum 75, 146
  var. *bulbilliferum* 75
  var. *viviparum* 75
collinum 95
commutatum 68, 150
compactum 146
condensatum 85
confertum 126
confusum 100
continuum 85
controversum 86, 128
convallarioides 86
coppoleri 117
coryi 60, 86–7, 93
cowanii 86, 87, 112
crenulatum 37, 52, 87, 145
crispum 87–8, 120
cupanii 88, 88, 150
  subsp. *anatolicum* 88
  subsp. *cupanii* 88
  subsp. *hirtovaginatum* 88
cupuliferum var. *regelii* 125
cuspidatum 63, 109
cuthbertii 88–9
cyaneum 36, 37, 38, 56, 58, 60, 71, 89, 89–90, 135, 152, 152
  var. *brachystemon* 135

cyathophorum 90
  var. *farreri* 37, 38, 53, 55, 58, 60, 72, 90, 90–1, 140, 147
cyrilli 91, 151
dalmaticum 100
darwasicum 23
decipiens 112, 151
delicatulum 109, 150
derderianum 91
descendens 67, 135
diaphanum 103
dichlamydeum 37, 55, 91–2, 92
dictyoprasum 60
dictyotum 98
dilatatum 150
dilutum 113
douglasii 16, 38, 92–3, 106
  var. *columbianum* 92
  var. *constrictum* 92
  var. *douglasii* 90, 92
  var. *nevii* 92
drummondii 78, 87, 93
eginense 84
elatum 36, 64, 93, 105
elburzense 72
equicaeleste 106
ericetorum 60, 93–4, 94, 148
euboicum 118
eurotophilum 58, 100
exsertum 83
falcifolium 45, 46, 55, 74, 94
  var. *breweri* 94
  var. *demissum* 94
fallax 134
farreri 59, 90, 91
favosum 150
fibrillum 94–5
fibrosum 98
filifolium 122
fimbriatum 95
  var. *abramsii* 95
  var. *denticulatum* 95
  var. *diabolense* 95
  var. *mohavense* 95
  var. *munzii* 95
  var. *parryi* 95
  var. *purdyi* 38, 95
  var. *sharsmithae* 95
fistulosum 24, 59, 60, 66, 95, 95–6, 97, 148
flavescens 65, 66
flavidum 136
flavum 53, 56, 58, 59, 60, 96–7, 97, 118, 121, 122, 150
  ex-Wisley 96
  pumilum roseum 97
  subsp. *flavum* 96
  subsp. *tauricum* 97
  subsp. *tauricum* var. *pilosum* 97
  var. *calabrum* 97

  var. *minus* 97
  var. *nanum* 96
  var. *pulchellum* 79, 123
  var. *tauricum* 97
flexum 79
'Forescate' 53, 62, 129, 152
forrestii 37, 38, 97
freynianum 122
freynii 122
frigidum 23, 109, 150
funiculosum 98
fuscum 118
gaditanum 100
gageanum 36
galanthum 59, 97
geyeri 15, 16, 16, 37, 98
  var. *geyeri* 98
  var. *tenerum* 98
giganteum 36, 57, 64, 93, 98, 99, 105, 117
glandulosum 60, 99
glaucum 133, 134
'Globemaster' 99, 99
globosum 128
gomphrenoides 150
goodingii 58, 99, 100, 100
graecum 141
grandiflorum 110
grandisceptrum 143
grosii 109, 150
guicciardii 97
guttatum 100–1, 150
  subsp. *dalmaticum* 100
  subsp. *guttatum* 100
  subsp. *sardoum* 100
haemanthoides 72
  var. *lanceolatum* 91
haematochiton 101
halleri 67
heldreichii 101, 150
helicophyllum 101
helleri 93
hendersonii 93
hierochuntinum 20, 69
hirsutum 138
hirtifolium 64, 101–2, 102
hirtovaginatum 88
hirtovaginum 150
horvatii 60, 148
hyalinum 55, 102
hymenorhizum, *hymenorrhizum* 102, 134, 148
hymettium 60, 150
incarnatum 126
inderiense 103, 124, 148
inequale 109, 150
insubricum 55, 56, 60, 103, 103–4, 110, 111, 148
integerrimum 150
jajlae 131
japonicum 140

jenischianum 113
jesdianum 64, 104
jubatum 150

kansuense 60, 71, 135
karataviense 25, **26**, 36, 55, 62, 65, **104**, 104–5, 107
kazerouni 104
kermesinum 134, 148
kharputense 32, **33**
kochii 146
komarovianum 140
kunthii 99
kurtzianum 115

lacteum 112
lancifolium 120, 147
latifolium 65
latissimum 145
ledebourianum 23
libani 23, 37
lineare 59, 62, 105, 105, 134, 135, 136, 148
longanum 149
longicaule 78
longicuspis 22, 127
longispathum 118
loscosii 135
lusitanicum 134
luteolum 60, 150

macedonicum 150
macleanii 64, 93, 105–6
macranthum 57, 59, 60, 106
macrorhizum 102
macrum 106
madidum 95
magicum 113
mairei 38, 49, 55, 56, 58, 66–7, 106–7, 147
majale 126
margaritaceum 100
  var. *guttatum* 100
marschallianum 128
marvinii 101
massaessylum 149
melanantherum 150
melananthum 150
meteoricum 109, 150
microdictyum 145
mirum 55, 56, 65, 91, 107, 107
moly 13, 13, 48, 48, 53, 60, 108, 131, 149
  minus 39
  'Jeannine' 53, 108
  var. *bulbilliferum* 108
monospermum 68
monspessulanum 113
montanum 134
montigenum 120
morrisonense 140
moschatum 108, 108–9, 150
multibulbosum 112, 113
murrayanum 109–10, 143

mutabile 78, 93
myrianthum 86

narcissiflorum 1, 55, 56, 103, 104, 110–11, 148
narcissifolium 110
neapolitanum 60, 62, 87, 111, 111–12, 149
nebrodense 97
neriniflorum 112
nerinifolium 112
nevii 93
nigrum 32, 60, 91, 112, 151
  var. *atropurpureum* 69
  var. *cyrilli* 91
niveum 138
noëanum 113
nutans 113
nuttallii 38, 93, 152

obliquum 24, 24, 57, 59, 60, 113–14, 148
obtusiflorum 88, 150
obtusifolium 80
occidentale 68
ochotense 145
ochroleucum 94
odorum 42, 124
oleraceum 23, 24, 35, **114**, 114, 150
olympicum 36, 55, 56, 115, 114–16
ophiopogon 140
ophioscorodon 86
oreophilum 36, **116**, 116
  var. *ostrowskianum* 53, **54**
  'Zwanenburg' 116
oreoprasum 116–17
orientale 151
oschaninii 20, 81, 117
ostrowskianum 53, 116
  'Zwanenburg' 116
oviflorum 106
oxphilum 82

paczoskianum 97
palentinum 134, 148
pallens 56, 86, 117, 127, 150
  subsp. *pallens* 117
  subsp. *siciliense* 117
  subsp. *tenuiflorum* 117
palmeri 72
paniculatum 56, 58, 59, 60, 74, 114, 117–18, 118, 127, 150
  subsp. *euboicum* 118
  subsp. *fuscum* 118
  subsp. *paniculatum* 118
  subsp. *villosulum* 118
  var. *legitimum* 118
  var. *macilentum* 127
  var. *pallens* 117, 127
  var. *rupestre* 127
  var. *tenuiflorum* 117
  var. *villosulum* 118

paradoxum 35, 59, 60, 119, 149
  var. *normale* 7, 55, 59, 60, 119
parciflorum 88, 150
pardoi 68, 150
parnassicum 150
'The Pearl' 112, 112
pedemontanum 111
pendulinum 55, 59, 60, 61, 119, 149
peninsulare 55, 62, 87, 120, 120
  var. *crispum* 88
permixtum 138
peroninianum 88
pervestitum 60, 150
phariense 84
phthioticum 149
pikeanum 98
pilosum 150
pisidicum 88
platyphyllum 120, 147
platystemon 116
plummerae 58, 100
podolicum 150
polyanthum 68, 150
polyastrum 37, 38, 57, 59, 60, 97, 120–1, 147, 152, **162**
  var. *platyphyllum* 147
polyphyllum 80
porrum 22, 67, 121
potaninii 124
praecissum 118
praecox 102
procerum 98
propontichum 150
protensum 55, 65, 121, 129
pruinatum 150
pseudocepa 97
pseudocyaneum 140
pseudoflavum 60, 121–2
pseudojaponicum 140
pskemense 122
pulchellum, (*carinatum* subsp. *pulchellum*) 60, 79, 122–3
purdomii 37, 89, 90
'Purple Sensation' 57, 123
purpurascens 129
purpureum 135
pyrenaicum 59, 68, 69, 86, 123, 150
pyrrhorrhizum 106, 147

raddeanum 129
ramosum 123–4, 142
recurvatum 82
reflexum 84
regelianum 150
regelii 124, 124–5
reticulatum 15, 24, 139
  var. *playanum* 139
rhizomatum 99
rhodopeum 118

rilaense 146
robinsonii 125, 125
rollii 67
rosenbachianum 36, 57, 64, 104, 125–6, 137
roseum 24, 35, 60, 126, 149
  subsp. *bulbiferum* 126
  var. *bulbiferum* 126
  var. *bulbilliferum* 126
  var. *carneum* 126
  var. *humile* 126
  var. *insulare* 126
  var. *roseum* 126
rotundum 24, 131
  subsp. *waldsteinii* 131
rouyi 88, 150
rubellum 25, 109, 150
rubens 126, 134, 136, 148
rubicundum 36
rubrovittatum 150, 151
rubrum 98
'Ruby Gem' 23
runyoni 93
rupestre 127, 127, 150
ruprechtii 128
rydbergii 98

sabulicola 98
sabulosum 150
sacculiferum 140
sardoum 100
sativum 22, 86, 127–8, 150
  subsp. *ophioscorodon* 86, 128
  var. *ophioscorodon* 86
savranicum 128
saxatile 128, 128, 134, 148
scaberrimum 68, 150
scabriscapum 124
scabrum 114
scaposum 23
schmitzii 148
schoenoprasum 23, 35, 53, 59, 62, 101, 129, 148
  'Forescate' 53, 54, 62, 129, 152
  var. *sibiricum* 59, 62
schubertii 55, 56, 74, 121, 129, 131
scorodoprasum 19, 35, 59, 60, 114, 131, 146, 150
  subsp. *jajlae* 130, 131
  subsp. *rotundum* 24, 70, 85, 131
  subsp. *scorodoprasum* 131
  subsp. *waldsteinii* 131
scorzonerifolium 131
  var. *scorzonerifolium* 60, 130, 131–2, 149
  var. *xericiense* 131
segetum 67

senescens 25, 68, 92, 128, 132, 133–4, 136, 137, 148, 152
  subsp. *montanum* 133, 133
  subsp. *senescens* 133
  var. *calcaneum* 134
  var. *glaucum* 133, **134**
serbicum 117
serratum var.
    dichlamydeum 92
setaceum 109
sibiricum 35, 129
siculum 75, 134
sieberianum 112
sikkimense 36, 49, 55, 56, 58, 60, 71, 89, **134**, 134–5, 140
simethis 106
simillimum 63
sipyleum 150
'Siskiyou Mt' ('Siskyou Mt' – Moore) 23
sp. AC&W 1956 115
sp. AC&W 2353 115
sp. AC&W 2354 115
sp. AC&W 2372 115
sp. AC&W 2375 115
sp. aff. *A.stamineum* 115
sp. JCA 361 55, 116
sp. Tibet 53, 59, 60
sphaerocephalon 25, 35, ·36, 53, 57, 135, 150
  subsp. *arvense* 135
  subsp. *sphaerocephalon* 135

subsp. *trachypus* 135
var. *descendens* 135
var. *trachypus* 135
var. *typicum* 135
var. *viridi-album* 135
sphaeropodum 97
spirale 133, 134
splendens 62, 135–6
  var. *kurilense* 136
spurium 134
stamineum 115, 150
staticiforme 150
stellatum 25, 37, 136
  'Album' 136
stelleranum 136, 148
stellerianum 126, 136
stenanthum 73
steveni var. β 128
stipitatum 36, 57, 64, 99, 102, 123, 126, 137
strictum 105
suaveolens 62, 69, 133, 137, 148
subhirsutum 62, 137–8, **138**, 141, **149**, 149
  subsp. *trifoliatum* 141
  var. *graecum* 141
  var. *subvillosum* 139
  var. *trifoliatum* 141
subquinqueflorum 127
subvillosum 138–9, 149
sulcatum 112
synnotii 36

talijevii 150
taquetii 140

tardens 150
tartaricum 124
tataricum 71, 103, 124
tauricum 97
tchongchanense 147
tenellum 72
tenuiflorum 117
textile 24, **139**, 139
thomsonii 80
thunbergii 37, 124, 136, 139–40, **140**
  'Ozawa' 140
  'Ozoke' 140
tibeticum 23, 53, 60, 90, 135, 140
tolmiei 120
trachypus 135
tricoccum 25, 25, 37, 59, 62, 140–1, 144
  var. *burdickii* 140
trifoliatum 141, 149
trilophostemon 79
triquetrum 27, 27, 35, 36, 55, 59, 60, 62, 119, 141, 149
triste 127
tristissimum 127
tuberosum 15, 24, 42, 57, 57, 59, 62, 123, 124, 142, 152

uliginosum 142
unifolium 37, 55, 62, 89, 109, 112f, **142**, 142–3
  var. *lacteum* 143

ursinum 18, 18, 25, 35, 36, 37, 47, 59, 62, 119, 140, 143–4, 150
  subsp. *ucrainicum* 144
  subs. *ursinum* 144

validum 58, 73, 100, **144**, 144–5
vancouverense 87, 145
victorialis 16, **17**, 36, 37, 140, 145, 149
  var. *listera* 145
vineale 25, 35, 47, 59, 60, 114, 131, 145, 150
  var. *capsuliferum* 146
  var. *compactum* 145
  var. *purpureum* 146
  var. *typicum* 145
  var. *virens* 146
virgunculae 37, 56, 146
viride 60
viviparum 75, 146

waldsteinii 131
wallichianum 23
wallichii 57, 59, 60, 120, 146–7
  var. *platyphyllum* 147
watsonii 87
webbii 97
weichanicum 124
wincklerianum 23

yatei 125
yunnanense 106, 147

zebdanense 11, 55, 62, 147